LOGIC-BASED 0-1 CONSTRAINT PROGRAMMING

OPERATIONS RESEARCH / COMPUTER SCIENCE INTERFACES SERIES

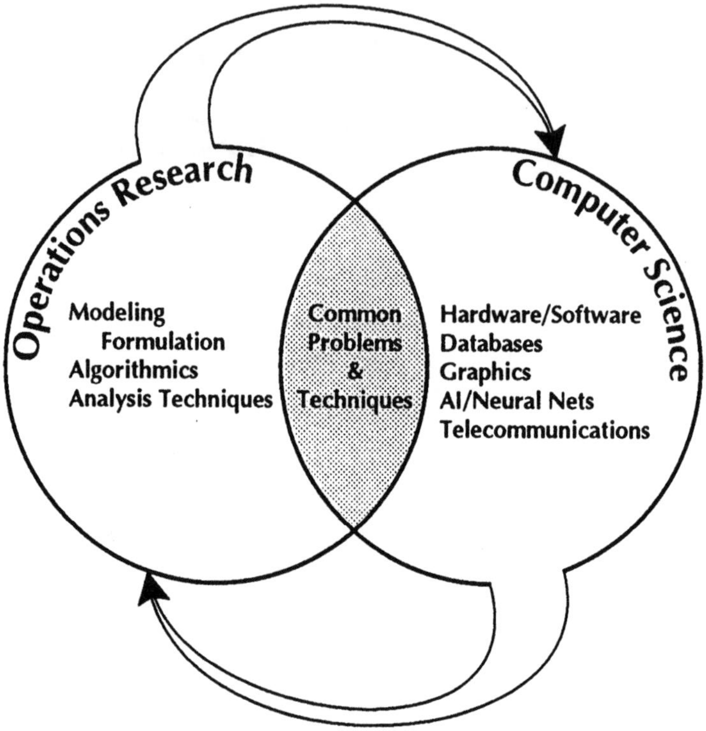

Series Editor

Ramesh Sharda
Conoco/Dupont Chair of Management of Technology
Oklahoma State University
Stillwater, Oklahoma U.S.A.

Other published titles in the series:

Brown, Donald/Scherer, William T.
University of Virginia
Intelligent Scheduling Systems

Nash, Stephen G./Sofer, Ariela
George Mason University
The Impact of Emerging Technologies on Computer Science and Operations Research

LOGIC-BASED 0-1 CONSTRAINT PROGRAMMING

Peter Barth
Max Planck Institute
Saarbrücken, Germany

KLUWER ACADEMIC PUBLISHERS
Boston/London/Dordrecht

Distributors for North America:
Kluwer Academic Publishers
101 Philip Drive
Assinippi Park
Norwell, Massachusetts 02061 USA

Distributors for all other countries:
Kluwer Academic Publishers Group
Distribution Centre
Post Office Box 322
3300 AH Dordrecht, THE NETHERLANDS

Library of Congress Cataloging-in-Publication Data

A C.I.P. Catalogue record for this book is available from the Library of Congress.

ISBN-13: 978-1-4612-8564-9 e-ISBN-13: 978-1-4613-1315-1
DOI: 10.1007/978-1-4613-1315-1

Printed on acid-free paper.

CONTENTS

FOREWORD

Constraint programming has become a major area of interaction between operations research and computer science over the last few years. It combines new programming language paradigms from computer science, like *constraint logic programming* and *concurrent constraint programming*, with efficient constraint solving techniques from mathematics, artificial intelligence, and operations research. Constraint programming allows the user both to build and to solve a model in the same framework:

- Expressive programming language constructs make it possible to formulate the problem in a natural and declarative way.
- Powerful constraint handling techniques supporting these language constructs provide for efficiency in problem solving.

Compared to other approaches, constraint programming offers incremental development, flexibility for modifications and extensions, and interactivity, without sacrificing efficiency. Constraint technology has been applied successfully to a large variety of practical problems arising in industry, for example in production planning, scheduling, and resource allocation.

0-1 constraint programming, where constraints are defined over 0-1 variables, corresponds to 0-1 integer programming in operations research. 0-1 constraints are a special form of finite domain constraints, which are used in most existing constraint programming systems and which are usually solved by local consistency techniques from artificial intelligence. In his book, Peter Barth develops a new *logic-based* approach for 0-1 constraint solving in constraint programming, which combines methods from 0-1 integer programming with techniques from automated deduction in computational logic. The basic formulas of his constraint solver are extended clauses saying that at least m out of n literals have to be true, where a literal is either a 0-1 variable or its negation. He has developed a comprehensive collection of new techniques for transforming general, possibly non-linear, 0-1 constraints into extended clauses, for simplifying sets of extended clauses by the generation of *logic cutting planes*, and for computing solved forms in the context of constraint programming. For the first time, a complete account of the techniques needed for a logic-based 0-1 constraint solver, or a logic-based 0-1 optimizer, is given.

This logic-based approach is complementary to the more classical polyhedral approach in combinatorial optimization, which is based on linear programming relaxations. Practical experience shows that logic-based methods can be very effective where polyhedral methods fail and vice versa. There are many interesting parallels between the two approaches, but the relationship between them is still not fully understood. This book makes a significant contribution to the area. It lays the foundation for future research, which has to show how logic-based and polyhedral methods can be combined in the most fruitful way.

Alexander Bockmayr

PREFACE

Logic-based methods for modelling and solving combinatorial problems have recently started to play a significant role in both theory and practice. The application of logic to combinatorial problems is twofold. On one hand, *constraint logic programming* allows one to declaratively model combinatorial problems over an appropriate constraint domain, which are then solved by a corresponding constraint solver. Besides being a high-level declarative interface to the constraint solver, the logic programming language allows one to implement also those subproblems that cannot be naturally expressed with constraints. On the other hand, logic-based methods can be used as a *constraint solving* technique inside a constraint solver for combinatorial problems modelled as 0-1 integer programs. A logic view of 0-1 integer programming problems provides new insights into the problem structure and may lead to more effective solving techniques depending on the problem class. The combination of operations research techniques and deductive methods provides effective constraint solving techniques from operations research integrated in a logic programming environment.

Pseudo-Boolean or 0-1 constraints are expressive enough to formulate a wide range of practical problems. We develop a set of techniques needed for building a logic-based constraint solver for 0-1 integer programming problems and demonstrate their integration into classical operations research methods. Logic-based methods are neither a priori superior to classical polyhedral techniques nor do they entail a completely different constraint solving methodology. Instead, there are many parallels to the polyhedral approach and achievements in one area carry over to the other. In polyhedral theory we can find the notion of a solved form, i.e., an inequality description of the convex hull of the 0-1 solution points of a pseudo-Boolean problem. We introduce an analogous logic-based solved form consisting of extended clauses. Extended clauses state that at least k out of n facts need to hold and hence are suitable for combinatorial problems. Lying halfway between expressive linear pseudo-Boolean constraints and manageable classical clauses (where $k = 1$), extended clauses are a good compromise w.r.t. expressibility and manageability. Since the possibly exponential size of the ideal solved form forbids its generation in general, the logic-based constraint solver works on a mixture of linear pseudo-Boolean inequalities and extended clauses. The techniques for extended clauses are expanded to cope with linear pseudo-Boolean inequalities and simplification methods for linear pseudo-Boolean inequalities are included. The logic-based methods can also co-operate with polyhedral methods. For instance, extended clauses generated by our transformation technique correspond to extended cover inequalities, which have been successfully used as cutting planes in linear programming based systems. Logic based simplification methods can yield even stronger polyhedral cuts. Ultimately, an integration of both approaches will yield more stable and efficient optimization tools.

Overview

We introduce constraint logic programming in Chapter 2. After a motivating example and a presentation of semantic issues and problems, we illustrate modelling techniques in CLP($\mathcal{PB}$), a constraint logic programming language over pseudo-Boolean constraints. We also discuss other constraint logic programming systems and constraint domains that are suitable for handling combinatorial problems. After reviewing previous work on pseudo-Boolean problems, we focus on a logic-based constraint solving approach in Chapter 3. An ideal logic-based solved form consisting of extended clauses in defined in Section 3.3. We present a complete logic-based pseudo-Boolean constraint solver in Chapter 4 and motivate each of the techniques developed in the following chapters. We start with a generalization of the unit resolution procedure for classical clauses to the pseudo-Boolean case in Chapter 5, which serves as a discrete relaxation replacing the classical linear programming relaxation. An implicit enumeration procedure based on this discrete relaxation is presented, which compares well with traditional linear programming based branch-and-bound. Next, we show in Chapter 6 how to use this enumerative method in order to generate valid constraints, i.e., logic cuts, replacing classical polyhedral cuts in a logic view of pseudo-Boolean constraints. Since extended clauses constitute the ideal solved form, we present techniques transforming linear pseudo-Boolean constraints into extended clauses in Chapter 7. Specialized simplification techniques for extended clauses are developed in Chapter 8, which are the logical counterparts of strong cutting planes in polyhedral theory. In order to handle non-linear pseudo-Boolean constraints we develop a linearization technique in Chapter 9. Finally, we present in Chapter 10 a projection technique in order to eliminate auxiliary variables, which is needed in the context of constraint logic programming.

Audience

The intended audience of this book can be divided into two groups. First, developers and users of constraint logic programming systems will be interested in using operations research techniques for combinatorial optimization problems. We discuss the requirements of an integration of such techniques in constraint logic programming, state upcoming problems, and present solutions for our logic-based approach. Since our constraint solver is able to decide entailment problems, the techniques are also relevant for designers of concurrent constraint languages. From a user's perspective, a complete constraint solver for 0-1 problems allows for a more declarative programming style and hence eases implementation.

Second, people interested in operations research will find new logic-based techniques for solving 0-1 problems. In some applications, exploiting the logical structure turns out to be superior to the classical polyhedral view underlying linear programming based systems. Software engineers faced to practical problems will be interested in applicable solving techniques. A prototype implementation of some of the presented techniques has been released as the system OPBDP [Bar95] and about hundred users successfully have applied the system to 0-1 problems coming from such diverse areas as statistics, VLSI-CAD, scheduling, and others. Researchers interested in the logic view of 0-1 constraints will find a well developed set of logic-based techniques, which can be used, refined, and may stimulate the development of new logic-based methods.

Acknowledgements

This book is a revised version of my PhD thesis, submitted at the University of Saarbrücken in February 1995. I am indebted to many people who made this work possible. First, I would like to thank my supervisors Prof. Dr. Harald Ganzinger and Dr. Alexander Bockmayr. Many thanks to Prof. Dr. Harald Ganzinger for his interest in my research and his support during my studies. He gave me the opportunity to work in an excellent research group at the Max Planck Institute in Saarbrücken. I am indebted to Dr. Alexander Bockmayr for his strong encouragement and constant support for this work. I took advantage of his valuable comments and many fruitful discussions. I thank him for his patience in reading and correcting numerous draft versions, which substantially improves this work. He introduced me to the problem area and kindly supported and advised me through my PhD studies. Special thanks to Prof. Dr. John N. Hooker for his instant commitment to become one of the referees, for his careful reading of the entire text, and for his helpful comments.

Finally, I am very grateful to my room-mates Manfred Jaeger and Peter Graf for the nice atmosphere and for spending time for discussions and reading draft versions.

1

INTRODUCTION

We present two orthogonal ways of using logic for modelling and solving combinatorial optimization problems. First, we demonstrate the use of *Constraint Logic Programming* (CLP) in order to model combinatorial problems and handle non-constraint aspects of real-life problems. Second, we propose and evaluate a *logic-based constraint solving* approach for pseudo-Boolean constraints, exemplify the potential of combining logic-based and polyhedral methods, and develop all the essential elements of a logic-based constraint solver for pseudo-Boolean constraints that can be used in the context of constraint logic programming.

1.1 CLP AND COMBINATORIAL OPTIMIZATION

Constraint logic programming combines the declarative nature of logic programming with the computational power of constraint solvers over specific domains. The logic programming language PROLOG, emerging from the work of Colmerauer [CKPR73] and Kowalski [Kow74], has introduced a subset of first-order logic, namely Horn clause logic, as a programming language. Programming in logic supports the writing of clear, readable, and concise programs. Although common constraint systems, like arithmetic, can be defined from scratch in logic, efficiency considerations forbid that. The CLP($\mathcal{X}$)-scheme of Jaffar and Lassez [JL87] merges logic programming and constraint solving, and thus provides a framework for both declarative and efficient programming. Given a constraint solver over a computational domain $\mathcal{X}$ that is able to decide the consistency of a set of constraints over $\mathcal{X}$, the CLP($\mathcal{X}$)-scheme provides a semantic basis for the corresponding constraint logic programming language CLP($\mathcal{X}$). Instantiating $\mathcal{X}$ by a domain of computation, such as linear arithmetic, Boolean algebra, finite domains, or lists, has been an active research area in recent years. Several constraint logic programming systems have been developed. Among the pioneering systems are CLP($\mathcal{R}$) [JL87], PROLOG III [Col87], CAL [ASS+88], and CHIP [DvHS+88a, vH89a].

Handling combinatorial optimization problems within a logic programming language is one of the aims of CHIP. With its finite domain constraint solver, based on local consistency techniques [Mac77, vH89a], the system has been successfully applied to a large variety of combinatorial optimization problems coming from operations research; e.g., graph coloring, disjunctive scheduling, car-sequencing, warehouse location, and cutting-stock problems [DvHS+88b, DSvH88b, DSvH88c, DSvH88a, DSvH90]. In a CLP-language over finite domain constraints with a constraint solver based on local consistency techniques, a combinatorial problem is viewed as a constraint satisfaction problem. The main task of such a language is the implementation of specific *constraint solvers* [vH89a]. Local consistency based solvers are also called incomplete constraint solvers, since they do not solve a combinatorial problem but a relaxation of the problem. For deciding global consistency of a set of constraints, the programmer has to rely on enumeration. Handling combinatorial problems with local consistency techniques inside a CLP-language has the following features:

Modelling: Inherited from the logic-programming paradigm the use of CLP allows for a declarative modelling of combinatorial optimization problems. Aspects of a real-life problem that are not (or not naturally) expressible as constraints, i.e., interactivity, incrementality, rule-based conditions, dynamic modifications, etc., can be handled by exploiting the programming language aspect of CLP. This tight coupling of constraint solving and programming offers a flexibility that favors CLP w.r.t. pure 'modelling languages' like GAMS [MB93] and AMPL [FGK90].

Constraint Solving: Embedding local consistency techniques, coming from AI, into logic programming offers a convenient environment for handling constraints (CHIP = Constraint Handling In Prolog) arising in combinatorial optimization problems. Search and backtracking, the main paradigms for solving constraint satisfaction problems with local consistency techniques, are naturally available in PROLOG. This yields a clean and straightforward integration. Constraint solvers based on the constraint satisfaction approach can be implemented in a logic programming language with integrated local consistency based finite domain constraint solving featuring a short development time and great flexibility.

A different approach for handling combinatorial problems inside logic programming is to embed a complete constraint solver for combinatorial problems ensuring global consistency. This enables us to model and program with combinatorial problems without caring about constraint solving and hence emphasizes the declarative modelling aspect. Non-constraint aspects of the problem then can be freely mixed with constraints without worrying about semantic problems arising in the obligatory enumeration phase of a local consistency based solver. Furthermore, in such a language the constraint solver can exploit any constraint solving technique available, and thus use achievements from the operations research community, and need not rely on local consistency techniques combined with enumeration. Disadvantages of the local consistency approach are discussed in Section 2.4. The features of a CLP-system offering constraint solving for combinatorial problems are presented in Section 2.5.2.

Pseudo-Boolean constraints have long been used for modelling combinatorial problems in the operations research community. They can be viewed as a restriction of finite domain constraints where each variable has the special domain $\{0,1\}$. Modelling with pseudo-Boolean constraints covers a large fraction of typical operations research problems like scheduling and sequencing [For59, For60, HR68, PWW69, Tah71], time-table scheduling, coding theory [HR68], plant location [Ham68], inter-city traffic [Rhy70], kinetic energy in spin glass models [KGV83], fault diagnosis [WYO81], cutting problems [CW77, Bea85], or supply support of space stations [FGGB66]. In mathematics and operations research solving optimization problems over pseudo-Boolean constraints ($\mathcal{PB}$) is also known as *constrained non-linear 0-1 programming* [HJM93]. We obtain the constraint logic programming language CLP($\mathcal{PB}$), introduced by Bockmayr [Boc91, Boc93b], when instantiating the computational domain $\mathcal{X}$ of the CLP($\mathcal{X}$)-scheme to pseudo-Boolean constraints.

Our aim is to design a complete constraint solver for pseudo-Boolean constraints that can be embedded in a logic programming language, i.e., in CLP($\mathcal{PB}$). The main differences w.r.t. a finite domain constraint solver based on local consistency techniques and our complete solver for pseudo-Boolean constraints are:

- *We always ensure global consistency.* Unlike local consistency based systems, we ensure at each point of the computation that the set of collected constraints is consistent. Hence, the CLP($\mathcal{X}$)-scheme gives us a clean semantics of our logic programming language.

- *We provide a solved form of the collected constraints.* A solved form of the collected constraints constitutes the output of a CLP-system. It should nicely describe the solution set of the modelled problem.
- *We support deciding entailment of constraints.* The solved form eases deciding entailment problems. The design of concurrent constraint logic programming (CCLP)languages necessitates a new functionality, namely deciding whether a constraint is entailed by a set of constraints, which we want to support. Additionally, entailment questions can be used for investigating a modelled problem.

In the context of constraint logic programming an appropriate solved form for finite domain constraints is not yet known. The restriction to the special domain $\{0, 1\}$ allows us to provide a constraint solver that fits the requirements of a constraint logic programming language, while a large fraction of combinatorial problems remains in the scope of the language. A CLP-language with a complete constraint solver encourages declarative programming and allows the use of constraints in order to guide the computation. A solved form that supports entailment allows a declarative reading of the solution set of the specified problem.

1.2 LOGIC-BASED PSEUDO-BOOLEAN CONSTRAINT SOLVING

Solving pseudo-Boolean constraints has a long tradition in the operations research community. In the last twenty years, most of the research in this field concentrated on linear programming based methods combined with branch-and-bound and polyhedral cutting plane techniques [NW88]. These methods have been very successful in solving some large instances of notoriously hard combinatorial problems, the most prominent example being the traveling salesman problem [PR91], by the development of polyhedral cutting plane techniques [Gom58, Chv73, CJP83, BCC94] yielding so-called branch-and-cut systems, e.g., MINTO [NSS94]. They have also been successfully applied to pure logical problems [Hoo88a]. On the other hand, logic-based methods for solving pseudo-Boolean problems have also been developed [HR68]. Interesting connections between cutting plane theory and logical inference have been established [Hoo88b, HF90, Hoo92b, Hoo94b, Hoo94a]. In this book we exploit a logic-based approach for solving pseudo-Boolean constraints in the context of constraint logic programming. We design an incremental logic-based constraint solver that is able to ensure global consistency and provides a solved form that eases deciding entailment, which meets the requirements of constraint logic programming.

In the late sixties, Hammer and Rudeanu [HR68] wrote a classical book on Boolean methods in operations research. Recently, there has been renewed interest in logic-based methods for combinatorial optimization. Hooker [Hoo94b] states that "Logic-based methods provide not only another point of view but an entire body of theory and algorithms that parallels the classical material. Linear relaxations, cutting planes, facet-defining cuts, convex hulls, etc. all have logical parallels". We elaborate on this connection and propose a logical view of pseudo-Boolean constraints that is based on the deductive system *Generalized Resolution* [Hoo92a] and fits into the framework of constraint logic programming.

Granot and Hammer [GH71] showed how to express pseudo-Boolean constraints by an equivalent set of classical clauses. Pseudo-Boolean constraints could then be solved with resolution-based methods [Rob65] in an incremental way [Jac92]. The solved form would then be a set of prime implicates[1], which fits exactly the requirements imposed by constraint solving in the context of constraint logic programming. Since a clause is entailed by a complete set of prime implicates if and only if it is implied by a single clause in that set, one of the main requirements is satisfied. The solved form makes deciding entailment trivial. Since a set of clauses is inconsistent if and only if the empty clause is a prime implicate, the solved form provides an easy check of consistency. So why are we not satisfied with that solution? The problem is that transforming arbitrary pseudo-Boolean constraints into an equivalent set of clauses typically generates an exponential number of clauses [GG80], which caused the failure of early attempts of solving pseudo-Boolean problems with clausal-based methods.

Recently, Hooker [Hoo92a] has presented a deductive system *Generalized Resolution* working with linear pseudo-Boolean inequalities and *extended clauses*, an extension of classical clauses. Semantically, a classical Boolean clause states that at least one of its literals has to be true. So a classical Boolean clause

$$L_1 \vee \ldots \vee L_n$$

can be expressed as linear pseudo-Boolean inequality

$$L_1 + \cdots + L_n \geq 1 \ ,$$

[1]The word "prime implicant" is used when we work with disjunctive normal form, opposed to "prime implicate" with conjunctive normal form.

where a negated literal $\overline{X_i}$ corresponds to $(1-X_i)$ and the truth values *false* and *true* correspond to 0 and 1 respectively. Extended clauses generalize classical clauses by allowing values other than 1 on the right-hand side. An extended clause is of the form

$$L_1 + \cdots + L_n \geq d \ ,$$

expressing that at least d of the n literals in the extended clause have to be 1. Hooker [Hoo92a] shows that there exists an equivalent notion of prime implicates for a set of extended clauses, and the deductive system *Generalized Resolution* generates such a set of *prime extended clauses.* Prime implicates and prime extended clauses are the logical counterparts to facet-defining cuts [NW88], the strongest polyhedral cuts.

We propose the use of extended clauses as basic constraints in our complete constraint solver for pseudo-Boolean constraints [Bar93a, Bar92b], because

- there is a method for computing a solved form and, given the solved form, deciding satisfiability and logical entailment is easy,
- the solved form describes all possible solutions and is explanatory, because of the natural d out of n interpretation,
- extended clauses are more expressive than classical clauses and provide a much more compact representation of information while preserving computational manageability.

For the latter point, note that the equivalent representation of the extended clause $L_1+\cdots+L_n \geq d$ using classical clauses without introducing new variables is

$$\bigwedge_{I\subseteq\{1,\ldots,n\}:|I|=n-d+1} (\vee_{i\in I} L_i) \ ,$$

a conjunction of $\binom{n}{n-d+1}$ classical clauses. The more compact representation often enables us to represent pseudo-Boolean constraints as an equivalent set of extended clauses, since the number of extended clauses needed is typically much smaller than the number of equivalent classical clauses.

When designing a pseudo-Boolean constraint solver based on extended clauses we are faced with two problems:

- Representing an arbitrary set of pseudo-Boolean constraints as an equivalent set of extended clauses.

- Computing the set of prime extended clauses equivalent to a set of extended clauses.

Both problems can be viewed as a reformulation of a set of constraints into an equivalent, but somehow simpler, set of constraints.

We break down the reformulation of pseudo-Boolean constraints into two parts. Given an arbitrary pseudo-Boolean constraint, we present reformulation techniques that allow us to

- linearize non-linear pseudo-Boolean constraints, i.e., reformulate an arbitrary pseudo-Boolean constraint as a set of linear pseudo-Boolean inequalities.
- transform a linear pseudo-Boolean inequality to an equivalent set of extended clauses.

Since there may be an exponential blow up in the size of the formulas when transforming them down to extended clauses, we present techniques that are sensitive to that blow up and only partially transform these formulas so that subsequent simplifications will make a full transformation easier. We also present a compact representation of pseudo-Boolean constraints that speed up the transformation process. In the context of constraint logic programming it is helpful to consider the set of already collected constraints while the transformation takes place. Thus, we can reduce the number of generated constraints and may generate immediately stronger constraints.

Powerful simplification of pseudo-Boolean constraints w.r.t. a set of pseudo-Boolean constraints is the key feature in order to process large pseudo-Boolean constraint sets. We present methods that strengthen linear pseudo-Boolean inequalities and extended clauses. We adapt the rules of the deductive system *Generalized Resolution* in order to generate *simplifying extended clauses.* The intention is that these extended clauses better approximate the set of prime extended clauses and thus approach the solved form. The simplification phase is the logical counterpart to the cut-generation phase in 0-1 integer programming. The simplifying extended clauses correspond to strong valid inequalities in polyhedral theory.

Pseudo-Boolean unit relaxation, a generalization of classical unit relaxation for linear pseudo-Boolean inequalities, is the logical counterpart to the linear programming relaxation in polyhedral theory. We present a logical branch-and-bound method for solving 0-1 integer programming methods based on a generalization of the well-known Davis-Putnam [DP60] enumeration algorithm for solving clausal satisfiability problems and show that this method is comparably efficient to state-of-the-art branch-and-bound systems based on linear programming.

In the context of constraint logic programming we favor a pure logic cut system, since there should be no branching inside the constraint solver. We show how to generate logic cuts by considering the pseudo-Boolean unit relaxation of the problem. In combination with simplification methods, this pure cut generation system will generate a complete set of prime extended clauses equivalent to an arbitrary set of pseudo-Boolean constraints.

The logic cut generation techniques and reformulation methods can also be applied inside linear programming based methods. We demonstrate this with several examples, i.e., preprocessing, coefficient reduction, extended cover generation, and simplification techniques are evaluated in the context of polyhedral methods. Moreover, these techniques can also be made available within a logic programming language suitable for implementing constraint solvers, like local consistency based systems whose solving approach can also be viewed as a logic-based branch-and-bound system. Then the generation of problem-specific cuts can be implemented and the development of logic-based branch-and-cut systems is supported on the language level.

Using logic-based techniques for solving pseudo-Boolean problems does not entail a radically different solution strategy; mainly important polyhedral concepts are replaced by their logical counterparts. Instead of the linear programming relaxation, a discrete relaxation is solved. Cutting plane generation techniques are replaced by the generation of logical implications. Modelling and solving experience collected by the operations research community can hence carry over to the logic-based approach. Furthermore, the logic view of pseudo-Boolean problems can be more natural than the polyhedral view. It turns out that exploiting the logic structure of a pseudo-Boolean problem is advantageous if the exploitation of the polyhedral structure seems to be inefficient or is just too complex. Logic-based methods already have found their way into polyhedral approaches. For example, the generation of cover inequalities, which are in fact logical implications of pseudo-Boolean constraints, has been widely recognized as an effective method in practice. Further exploration of logic-based methods and their combination with polyhedral approaches can yield a large contribution to the solution of pseudo-Boolean problems.

2

CONSTRAINT LOGIC PROGRAMMING

The declarative paradigms *constraint solving* and *logic programming* are successfully merged in constraint logic programming. Integrating constraint solving techniques into logic programming increases expressibility, efficiency, and flexibility and therefore encourages declarative programming. Efficient special purpose algorithms, coming from mathematics, operations research, or artificial intelligence, can be embedded into the logic programming language. The logic programming language allows one to declaratively state the problem, which is solved by the underlying constraint solver. Besides being a high-level declarative interface to the constraint solver, the logic programming language allows the implementation of subproblems that do not fit into the constraint domain.

First, we introduce the basic concepts of constraint logic programming by an example in Section 2.1. A more formal definition, which can be skipped on a first reading, is given in Section 2.2. We extract the special functionality of a constraint solver in the context of constraint logic programming in Section 2.3. The local consistency approach for solving combinatorial problems embedded in logic programming is compared with complete constraint solving inside CLP in Section 2.4. We introduce the constraint logic programming language CLP($\mathcal{PB}$), which handles combinatorial problems using pseudo-Boolean constraints in Section 2.5, and discuss programming examples for modelling typical practical problems. Finally, we review and compare existing logic programming systems for handling combinatorial problems in Section 2.6.

2.1 CLP BY EXAMPLE

We illustrate some crucial CLP-concepts by an example. For a detailed introduction in PROLOG we refer to [CM87]. In the program-fragments we use an *italic* font for constraints over $\mathcal{X}$. Here, $\mathcal{X} = \mathcal{PB}$, i.e., finite domain variables restricted to the domain {0,1} and constraints stated over these variables. For the rest of the code we use a `typewriter` font. Variables begin with an uppercase character.

Programming in a CLP($\mathcal{X}$)-language means declaring *facts* and *rules* about objects and their relationship, including constraints over $\mathcal{X}$. Facts and rules are also called *predicates*. Suppose we want to declare a relation `sequence` that holds for all lists of Boolean values where no consecutive chain of 3 zeros is allowed. A list containing the elements $X_1, X_2, \ldots, X_n$ is denoted by `[`$X_1, X_2, \ldots, X_n$`]`. Hence, the relation `sequence` consists of all lists

$$\{[X_1, X_2, \ldots, X_n] \mid X_i + X_{i+1} + X_{i+2} \geq 1 \text{ for all } 1 \leq i \leq n-2\} ,$$

where the X_i are Boolean variables and $X_i + X_{i+1} + X_{i+2} \geq 1$ is a pseudo-Boolean constraint.

A CLP-program that defines the relation `sequence` consists of 3 facts and one rule. First, we state the facts that the empty list and the one- and two-element-lists are `sequences`.

```
sequence([]).
sequence([X]).
sequence([X,Y]).
```

For lists that have more than two elements we define a recursive rule modelling the `sequence` property[1].

```
sequence([X,Y,Z|L]) :-
        X + Y + Z ≥ 1,
        sequence([Y,Z|L]).
```

The rule can be read as "A list containing at least three elements, where X, Y, Z are the first, second, and third elements respectively and `L` is the remaining list,

[1]The term `[`$X_1, X_2, \ldots, X_k$`|L]` denotes the first k elements of a list by $X_1, X_2, \ldots, X_k$ and the remaining list by `L`.

is a **sequence** if $X + Y + Z \geq 1$ holds and the list without the first element, i.e., $[Y, Z \mid L]$, is a **sequence**".

Once we have a CLP-program consisting of facts and rules we can ask *queries* w.r.t. this program. For example, we might be interested in knowing whether the list $[A, A, B, B, C, C]$ is a **sequence**. The query

?-**sequence**$([A, A, B, B, C, C])$.

gives rise to the answer

$A + B \geq 1,$
$B + C \geq 1$
yes.

This can be read as "The list $[A, A, B, B, C, C]$ is a **sequence** if $A + B \geq 1$ and $B + C \geq 1$ holds". Hence, in CLP, constraints constitute the output of a program. Operationally, this query is processed as follows. We search for a fact or a rule applicable to the query **sequence**$([A, A, B, B, C, C])$. If there are several applicable rules or facts these possibilities are tried on backtracking. Since the list $[A, A, B, B, C, C]$ contains more than two elements none of the facts apply, but the rule does. If $X = A, Y = A, Z = B$, and $L = [B, C, C]$, then the rule says that $[A, A, B, B, C, C]$ is a **sequence** if $A + A + B \geq 1$ and $[A, B, B, C, C]$ is a **sequence**. In other words, we replace the initial query by

$$A + A + B \geq 1, \mathbf{sequence}([A, B, B, C, C]) \ .$$

Computation now continues until only constraints remain or no rule or fact is applicable in which case *backtracking* takes place. Note that in order to apply the rule we have generated some equality constraints, which equalize the query and the left-hand side of the rule. The equality constraints together with the pseudo-Boolean constraints of the rule are added to the so-called *constraint store*. A rule or fact is only applicable if the constraint store remains consistent after adding the generated constraints, that is, the constraint store must admit a least one solution. Since rules and facts can be applied several times, all variables in a rule or fact are replaced by fresh ones at each invocation. We model this by subscripting the variables in the rules and facts by new indices whenever they are applied. For illustrative purposes we give the complete computation sequence of the previous query, i.e., the state of the constraint store and the remaining queries after a rule or fact has been applied.

sequence($[A,A,B,B,C,C]$).

$A = X_1, A = Y_1, B = Z_1, X_1 + Y_1 + Z_1 \geq 1,$
sequence($[A,B,B,C,C]$).

$A = X_2, B = Y_2, B = Z_2, X_2 + Y_2 + Z_2 \geq 1,\ A = X_1, A = Y_1, B = Z_1, X_1 + Y_1 + Z_1 \geq 1,$
sequence($[B,B,C,C]$).

$B = X_3, B = Y_3, C = Z_3, X_3 + Y_3 + Z_3 \geq 1,\ A = X_2, B = Y_2, B = Z_2, X_2 + Y_2 + Z_2 \geq 1,\ A = X_1, A = Y_1, B = Z_1, X_1 + Y_1 + Z_1 \geq 1,$
sequence($[B,C,C]$).

$B = X_4, C = Y_4, C = Z_4, X_4 + Y_4 + Z_4 \geq 1,\ B = X_3, B = Y_3, C = Z_3, X_3 + Y_3 + Z_3 \geq 1,\ A = X_2, B = Y_2, B = Z_2, X_2 + Y_2 + Z_2 \geq 1,\ A = X_1, A = Y_1, B = Z_1, X_1 + Y_1 + Z_1 \geq 1,$
sequence($[C,C]$).

$B = X_4, C = Y_4, C = Z_4, X_4 + Y_4 + Z_4 \geq 1,\ B = X_3, B = Y_3, C = Z_3, X_3 + Y_3 + Z_3 \geq 1,\ A = X_2, B = Y_2, B = Z_2, X_2 + Y_2 + Z_2 \geq 1,\ A = X_1, A = Y_1, B = Z_1, X_1 + Y_1 + Z_1 \geq 1.$

For all but the last step the rule has been applied. By replacing the X_i, Y_i and Z_i with their defining equalities, we see that the last state is equivalent to

$B + C + C \geq 1, B + B + C \geq 1, A + B + B \geq 1, A + A + B \geq 1,\ B = X_4, C = Y_4, C = Z_4,$
$B = X_3, B = Y_3, C = Z_3,\ A = X_2, B = Y_2, B = Z_2, A = X_1, A = Y_1, B = Z_1.$

Note that both $B+C+C \geq 1$ and $B+B+C \geq 1$ are equivalent to $B+C \geq 1$ and both $A+B+B \geq 1$ and $A+A+B \geq 1$ are equivalent to $A+B \geq 1$. Boolean variables that do not occur in the query should not occur in the answer. The auxiliary Boolean variables introduced by the execution of the CLP-program are projected away and the answer constraint $A + B \geq 1, B + C \geq 1$ is hence implied (and not necessarily equivalent) to the constraints in the final state.

Next, we trace the execution of a query that has no solution.

?- $A + C \leq 1$, sequence($[A,A,0,0,C,C]$).

The computed answer is no. We start with a similar sequence of states, but the fourth computation state is then

$0 = X_3, 0 = Y_3, C = Z_3, X_3 + Y_3 + Z_3 \geq 1,\ A = X_2, 0 = Y_2, 0 = Z_2, X_2 + Y_2 + Z_2 \geq 1,\ A = X_1, A = Y_1, 0 = Z_1, X_1 + Y_1 + Z_1 \geq 1,\ A + C \leq 1,$
sequence$([0, C, C])$.

which is equivalent to

$0 + 0 + C \geq 1, A + 0 + 0 \geq 1, A + A + 0 \geq 1,\ 0 = X_3, 0 = Y_3, C = Z_3, A = X_2, 0 = Y_2, 0 = Z_2, A = X_1,\ A = Y_1, 0 = Z_1,\ A + C \leq 1,$
sequence$([0, C, C])$.

Since $0{+}0{+}C \geq 1$ implies $C = 1$ and $A{+}A{+}0 \geq 1$ implies $A = 1$, we know that $A{+}C \leq 1$ is violated. Therefore, the constraint set is unsatisfiable, and in CLP no further computation is allowed. Since no other rule or fact is applicable on backtracking, the query fails and yields the answer no. This means that whatever the Boolean values of A and C are, the list $[A, A, 0, 0, C, C]$ is not a **sequence** if $A + C \leq 1$.

Finally, we show queries with their answers that nicely exemplify the descriptive power of constraints constituting the answer. The query

$$?\text{-}\mathbf{sequence}([A, B, \overline{A}, \overline{B}, A]).$$

where $\overline{X}$, the negation of a Boolean value, is an abbreviation for $1 - X$, yields the answer **yes** with no constraints imposed. From this we can conclude that *every* five element list of Boolean values where the successor of the successor is the negation of itself is a **sequence** regardless of the actual Boolean values. As we have seen not only simple Boolean variables can be used in a query but also more complicated pseudo-Boolean expressions. Consider for example the query

$$?\text{-}\mathbf{sequence}([A \cdot B, \overline{A} \cdot B, C \cdot A, A + B, D + \overline{A}]).$$

Note that also non-linear expressions are used. By applying the rule of our CLP-program the first time, we dynamically create and add the Boolean equations

$$\begin{aligned} A \cdot B &= X_1 \\ \overline{A} \cdot B &= Y_1 \\ C \cdot A &= Z_1 \end{aligned}$$

together with the constraint $X_1 + Y_1 + Z_1 \geq 1$. These constraints are checked for satisfiability and computation then continues. At the end of computation and after projecting away auxiliary Boolean variables we obtain the answer

$A + B \geq 1,$
$B + C \geq 1$
`yes.`

as for the query `?-sequence`$([A, A, B, B, C, C])$. This answer gives a nice and concise description of a complicated dynamically created pseudo-Boolean problem. Furthermore, since the Boolean variable D does not occur in the answer constraint, we know that the property of being a sequence does not depend on the value of D. Hence, constraint logic programming allows one to model, solve and investigate pseudo-Boolean problems. We have the following key features of constraint logic programming [JM94]:

- Constraints are used to specify the query as well as the answers.
- During execution, new variables and constraints are created.
- The collection of constraints in every state is tested as a whole for satisfiability before execution proceeds further and thus allows for control of execution.

2.2 SEMANTICS

The declarative and operational semantics of a CLP($\mathcal{X}$)-language is based on the semantics of pure logic programs, but parameterized by a choice of $\mathcal{X}$, the domain of computation and constraints. Let $\mathcal{X}$ be a 4-tuple $(\Sigma, \mathfrak{D}, \mathfrak{L}, \mathfrak{T})$, where Σ is a signature, $\mathfrak{D}$ is a Σ-structure, and $\mathfrak{T}$ is a first order Σ-theory [JM94]. The domain of computation is $\mathfrak{D}$, over which constraints of a class $\mathfrak{L}$ of Σ-formulas can be expressed, and $\mathfrak{T}$ axiomatizes properties of $\mathfrak{D}$. We assume that Σ contains the binary predicate $=$, which is interpreted as identity in $\mathfrak{D}$.

A *constraint domain* $(\mathfrak{D},\mathfrak{L})$ is *solution compact* [JM94] if every element in $\mathfrak{D}$ can be defined by a conjunction of constraints of $\mathfrak{L}$ and if the negation of each constraint of $\mathfrak{L}$ can be defined by a disjunction of constraints of $\mathfrak{L}$. The corresponding theory $\mathfrak{T}$ is *satisfaction complete* [JM94] w.r.t. $\mathfrak{L}$ if either $\mathfrak{T} \models \exists C$ or

$\mathfrak{X} \models \not\exists C$ for all $C \in \mathfrak{L}$. A solution compact constraint domain with a satisfaction complete theory constitutes an appropriate framework for a CLP($\mathcal{X}$)-language.

A CLP($\mathcal{X}$)-program consists of a finite set of constraint rules of the form

$$\mathtt{h} \text{ :- } \mathtt{b}_1, \ldots, \mathtt{b}_n \quad ,$$

where $\mathtt{h}$ is an atom and the $\mathtt{b}_i$ are either atoms over a signature Σ', not sharing any function or predicate symbols with Σ, or a constraint of $\mathfrak{L}$. A CLP($\mathcal{X}$)-goal is of the form

$$\text{?-}\mathtt{b}_1, \ldots, \mathtt{b}_n \quad .$$

By the results of Jaffar and Lassez [JL87, JM94], we have the existence of a canonical domain of computation, the existence of a least and greatest model semantics, the existence of a least and greatest fixed point semantics, soundness and completeness results for successful derivation of the underlying implementation model, soundness and completeness results for finitely failed derivations of the underlying implementation model, and soundness and completeness results for negation-as-failure.

An operational model [JM94] that describes the operational semantics of many CLP-systems is given by a transition system on *states*, i.e., tuples $\langle A, C\rangle$, where A is a multiset of atoms and C is a multiset of constraints[2], also called *constraint store*. The special state $\bot$ denotes failure. Intuitively, A is the set of not yet considered goals and C is the set of constraints collected so far. We assume to have a *computation rule* that selects an element of A and an appropriate transition rule for each state. The transition system is parameterized by a predicate consistent and a function infer corresponding to the specific constraint solver. An initial CLP-goal ?-$\mathtt{b}_1, \ldots, \mathtt{b}_n$ is represented by the state $\langle\{\mathtt{b}_1, \ldots, \mathtt{b}_n\}, \emptyset\rangle$. Suppose that the computation rule selects a constraint c in A, then rewriting the state $\langle A, C\rangle$ is defined by

$$\langle A, C\rangle \rightarrow \begin{cases} \langle A \setminus \{c\}, C'\rangle & \text{if} \quad \begin{array}{l} \mathsf{consistent}(C \cup \{c\}) \\ C' = \mathsf{infer}(C \cup \{c\}) \end{array} \\ \bot & \text{if} \quad \neg\mathsf{consistent}(C \cup \{c\}) \end{cases} .$$

Suppose that the computation rule selects an atom $\mathtt{a}$ in A and a constraint rule $\mathtt{h}$:- $\mathtt{b}_1, \ldots, \mathtt{b}_n$ of the constraint logic program such that $\mathtt{a}$ and $\mathtt{h}$ have the

[2]Note that C also contains equality constraints over finite trees over the signature Σ', introduced by unification when constraint rules are applied.

same predicate symbol, then

$$\langle A, C\rangle \rightarrow \begin{cases} \langle (A \setminus \{\mathtt{a}\}) \cup \{\mathtt{b}_1, \ldots, \mathtt{b}_n\}, C'\rangle & \text{if} \quad \begin{array}{c}\mathsf{consistent}(C \cup \{\mathtt{h} = \mathtt{a}\}) \\ C' = \mathsf{infer}(C \cup \{\mathtt{h} = \mathtt{a}\})\end{array} \\ \bot & \text{if} \quad \neg\mathsf{consistent}(C \cup \{\mathtt{h} = \mathtt{a}\}) \end{cases},$$

where $\mathtt{h} = \mathtt{a}$ is an abbreviation for the conjunction of equations between corresponding arguments of $\mathtt{a}$ and $\mathtt{h}$, extends the transition system $\rightarrow$. If there is an atom $\mathtt{a}$ in A, but no constraint rule in the CLP-program having the same predicate symbol, we have a failure.

$$\langle A, C\rangle \rightarrow \bot$$

The predicate consistent checks for consistency of a set of constraints C and is defined by

$$\mathsf{consistent}(C) \quad \text{if and only if} \quad \mathfrak{D} \models \exists \vec{x}\colon C$$

where $\vec{x}$ is the set of free variables in C. The function infer is responsible for simplifying a set of constraints C. We require that

$$\mathfrak{D} \models \mathsf{infer}(C) \Leftrightarrow C$$

and so preserve equivalence of the simplified constrained set. Jaffar and Maher [JM94] present a more sophisticated transition system that allows the distinction between active and passive constraints and the use of an incomplete consistency check. Furthermore, the infer function is split and a separate simplification transition rule is introduced. The presented operational semantics of a CLP-system described by $\rightarrow$ is called "ideal" in [JM94].

A *derivation* is a finite or infinite sequence of transition states

$$\langle A_0, C_0\rangle \rightarrow \cdots \rightarrow \langle A_i, C_i\rangle \rightarrow \cdots .$$

A *final state* is a state that cannot be rewritten by $\rightarrow$. A derivation is *successful* if its final state is of the form $\langle \emptyset, C\rangle$. Let G be a goal with free variables $\vec{x}$ that gives rise to a successful derivation with final state $\langle \emptyset, C\rangle$. Then $\exists \vec{x} : C$ is the *answer constraint* of this derivation. A derivation is *failed* if the final state is $\bot$. A derivation is *fair* if it is failed or if for all i each $\mathtt{a} \in A_i$ is rewritten in a later transition. A computation rule is fair if it generates only fair derivations. A CLP-goal is finitely failed if all derivations from the state representing the goal are failed.

The *computation tree* of a goal for a CLP-program in a CLP-system is a tree with nodes labeled by states, where the state of a direct subnode of a node is the state of the node rewritten once by $\rightarrow$. The root node of the tree has as label the state of the initial CLP-goal. In the computation tree each possible derivation for each possible computation rule is a branch of the computation tree. Executing a CLP-program, i.e., finding the answers of a CLP-goal w.r.t. a CLP-program, means searching all non-failed finite derivations in the computation tree and returning their answer constraint.

Most CLP-systems employ a standard depth-first search, which is not fair. The soundness and completeness results for an "ideal" CLP-system can be found in [JM94].

2.3 CONSTRAINT SOLVING

Let us investigate the domain specific predicate consistent and the function infer. We neglect the equality constraints over finite trees[3]. Along a derivation we have to perform a sequence of consistency checks, where at each step a new set of constraints is incrementally added to the previously consistent constraint set. For reasons of efficiency, it is advantageous to use information from the previous consistency checks in the sequence of current consistency checks. Simplification of the current constraint set by the function infer is one of the possible ways to achieve incrementality.

We again consider finite domain constraints over the domain $\{0, 1\}$ and follow a derivation of the goal

$$?\text{-}\mathtt{b}_1(A, B, C), \mathtt{b}_2(A, B, C), \mathtt{b}_3(A, B, C)$$

w.r.t. the CLP-program consisting of the three constraint rules

$$\begin{array}{ll} \mathtt{b}_1(A, B, C) & :\text{-}\ A + B \geq 1 \\ \mathtt{b}_2(A, B, C) & :\text{-}\ A + (1 - B) \geq 1 \\ \mathtt{b}_3(A, B, C) & :\text{-}\ A + C \geq 1 \quad . \end{array}$$

[3]Strictly speaking, a CLP($\mathcal{X}$)-language is a statically typed language, i.e., each argument in an atom and each variable must be used in a consistent way. Here, we also neglect the type problems and assume the presence of only well typed CLP($\mathcal{X}$)-programs [HJM+92].

First, assume that infer(C) $= C$ for all constraints C. Then a possible finite derivation sequence[4] is

$$\begin{aligned}&\langle\{A+B\geq 1, A+(1-B)\geq 1, A+C\geq 1\},\emptyset\rangle \rightarrow\\&\langle\{A+(1-B)\geq 1, A+C\geq 1\},\{A+B\geq 1\}\rangle \rightarrow\\&\langle\{A+C\geq 1\},\{A+B\geq 1, A+(1-B)\geq 1\}\rangle \rightarrow\\&\langle\emptyset,\{A+B\geq 1, A+(1-B)\geq 1, A+C\geq 1\}\rangle\end{aligned}$$

since the constraint sets of all states are consistent. The satisfiability check performed by consistent has to recheck satisfiability from scratch 4 times. Next, suppose that we have an infer that is more clever, i.e., infer is able to detect that $\{A+B\geq 1, A+(1-B)\geq 1\}$ is equivalent to $\{A=1\}$. Then the derivation sequence is

$$\begin{aligned}&\langle\{A+B\geq 1, A+(1-B)\geq 1, A+C\geq 1\},\emptyset\rangle \rightarrow\\&\langle\{A+(1-B)\geq 1, A+C\geq 1\},\{A+B\geq 1\}\rangle \rightarrow\\&\langle\{A+C\geq 1\},\{A=1\}\rangle \rightarrow\\&\langle\emptyset,\{A=1, A+C\geq 1\}\rangle\ .\end{aligned}$$

and checking consistency of the fourth state is much simpler. Note that the constraint set of the third state entails the newly added constraint of the constraint set of the fourth state. Therefore, consistency is trivially ensured and the new constraint is not needed. Given an infer[5] that eliminates entailed constraints, we obtain the derivation sequence

$$\begin{aligned}&\langle\{A+B\geq 1, A+(1-B)\geq 1, A+C\geq 1\},\emptyset\rangle \rightarrow\\&\langle\{A+(1-B)\geq 1, A+C\geq 1\},\{A+B\geq 1\}\rangle \rightarrow\\&\langle\{A+C\geq 1\},\{A=1\}\rangle \rightarrow\\&\langle\emptyset,\{A=1\}\rangle\ .\end{aligned}$$

Therefore, an *incremental check of consistency* for a set of constraints and *deciding entailment* of a constraint w.r.t. a set constraints are the basic operations to be provided in an implementation of consistent and infer.

Deciding entailment of a set of constraints is a basic functionality of a constraint solver in CCLP-systems [Sar89, HJ90, AK93, SHW93a]. This functionality is used for the synchronization of concurrent processes. We refer to [JM94, PS95] for an overview of CCLP-languages and the requirements on a constraint solver for CCLP-systems.

[4] For the sake of simplicity, we neglect the introduction of new variables like $A_i = A$.

[5] In an implementation infer and consistent are not strictly separated, but work hand in hand, e.g., it may be useful to eliminate entailed constraints before checking satisfiability.

Projecting a set of constraints onto a set of variables in order to eliminate extra variables not occurring in the initial CLP-goal has to be done at the end of computation in order to present a user-friendly answer constraint. Additionally, it may be useful to immediately project away variables that do not occur in the initial CLP-goal and that are no longer visible in the computation of the following states, i.e., no new constraints can be added involving these variables.

When we abstract away the inference machine from the operational semantics, we see that the following operations provide the basis for an implementation for a constraint solver over a specific domain [JM94].

consistency: Given a set of constraints $C \in \mathfrak{L}$, decide whether the set is consistent, i.e., whether $\mathfrak{D} \models \exists \vec{x} : C$, where $\vec{x}$ is the set of free variables in C.

entailment: Given a set of constraints $C \in \mathfrak{L}$, decide whether a constraint $C' \in \mathfrak{L}$ is entailed by C, i.e., whether $\mathfrak{D} \models \forall \vec{x} : C \Rightarrow C'$, where $\vec{x}$ is the set of free variables in C and C'.

projection: Let C be a set of constraints in $\mathfrak{L}$ and $\vec{x}$ be the set of free variables in C. Let $\vec{y} \subseteq \vec{x}$ be a set of variables that we want to project away. The problem is to calculate a simplified form C' of the constraint $\exists \vec{y} : C$. We require that $\mathfrak{D} \models C \Rightarrow C'$.

Instantiating the computational domain $\mathcal{X}$ of the CLP($\mathcal{X}$)-scheme and providing a constraint solver for the specific selected constraint domain has been an active research area in recent years. Before we formally instantiate the computational domain $\mathcal{X}$ of the CLP($\mathcal{X}$)-scheme to pseudo-Boolean constraints in Section 2.5, we will first point out some conceptual differences and semantic as well as practical problems that arise when viewing a logic-programming language with a local consistency based solver as a CLP-*language* over finite domain constraints.

2.4 LOCAL CONSISTENCY AND CLP

Logic programming languages with a local consistency based solver for finite domain constraints, like CHIP, is commonly viewed as a constraint logic programming language over finite domain constraints. However, contrary to common belief, this is not the case. As already mentioned, such a language is intended to ease the *implementation of user-defined constraint solvers using constraint satisfaction techniques* [vH89a] and therefore has a different intention than a CLP-language offering complete constraint solving of combinatorial problems. In the following we will demonstrate with examples what happens if one views a logic programming language with local consistency methods as CLP-language over finite domain constraints. We point out the pitfalls and problems and also motivate the need for a real CLP-language for modelling and solving combinatorial problems.

2.4.1 Local Consistency

Since combinatorial problems fall into the class of NP-complete problems, van Hentenryck [vH89a] concludes that "backtracking is an important technique for solving these problems". Since backtracking is naturally available in PROLOG, in CHIP the "most important operation on constraints ... a test for consistency" [JM94] of the current set of finite domain constraints is relaxed to local consistency and global consistency is no longer ensured. Instead, the programmer has to ensure global consistency by an extra enumeration predicate. A typical CLP-program for a system based on local consistency techniques therefore has the form

```
problem(<Variables>) :-
        <state constraints over Variables>,
        <enumerate domain of Variables>.
```

The search for a solution of the collected constraints is carried out by the PROLOG backtracking mechanism and the finite domain constraint solver is responsible for pruning the search tree. Logically, there is no need for an extra enumeration predicate. Given an "ideal" [JM94] constraint solver for finite domain constraints, a sufficient formulation of a CLP-program over finite domain constraints is

```
problem(<Variables>) :-
        <state constraints over Variables>.
```

For this kind of CLP-programs, a finite domain solver based on local consistency techniques will not generate a concise and simplified solved form. Consider the following query and assume that all variables are restricted to the domain $\{0, 1\}$.

$$?\text{-}A + B \geq 1, A + (1 - B) \geq 1.$$

The answer of a constraint logic programming language based on local consistency techniques (here we use ECLIPSE [MAC+89], which has a similar finite domain constraint solver as CHIP) is

`Delayed goals:` $A + B \geq 1, A - B \geq 0$ `yes.`

which is correct, since the constraint set is satisfiable, but not what we expect from a CLP-system. By adding an extra enumeration predicate to the previous query

$$?\text{-}A + B \geq 1, A + (1 - B) \geq 1, \mathbf{enum}([A, B]).$$

we obtain the following solutions from ECLIPSE.

$A = 1, B = 0$ `More? (;) ;`
$A = 1, B = 1$ `yes.`

By investigating the two solutions, we see immediately that the variable A has to be 1 for all possible solutions and that the variable B is arbitrary. Thus, a *good* answer of a CLP-system to the query

$$?\text{-}A + B \geq 1, A + (1 - B) \geq 1.$$

is

$A = 1$ `yes.`

The behavior is worse if the set of collected constraints is unsatisfiable. Assume again that all variables are restricted to the domain $\{0, 1\}$ and consider the following query.

$$?\text{-}A + B \geq 1, A + 1 - B \geq 1, 1 - A + B \geq 1, 1 - A + 1 - B \geq 1.$$

The answer of ECLIPSE,

$$\texttt{Delayed goals:}\ A+B \geq 1, A-B \geq 0, B-A \geq 0, 1-A-B \geq 0 \quad \texttt{yes.}$$

in particular, the yes is wrong, which can be verified by adding an extra enumeration predicate.

?- $A+B \geq 1, A+1-B \geq 1, 1-A+B \geq 1, 1-A+1-B \geq 1,$
   enum($[A, B]$).
no.

Unfortunately, adding an extra enumeration predicate does not ensure detection of failure in practice. Consider the following CLP-program over finite domain constraints, where the domain of the variables is restricted to $\{0, 1\}$.

f($[A, B]$) :-
      $A+B \geq 1, A+1-B \geq 1, 1-A+B \geq 1, 1-A+1-B \geq 1,$
      f($[A, B]$).

In ECLIPSE the query

?-f($[A, B]$), enum($[A, B]$).

goes into an infinite loop. If the underlying constraint solver ensures global consistency of the collected constraints, then the computed correct answer is no.

2.4.2 Global Consistency

Since the problems we consider are NP-complete, in the worst case assuring global consistency may require an exponential computation time in problem size (unless P = NP). Therefore, it has been argued that ensuring global consistency in the context of constraint logic programming is not efficiently achievable.

It is obvious that a CLP-program over finite domain constraints with a constraint solver based on local consistency techniques has to have an extra enumeration predicate in order to produce correct results. The advantage of delaying constraint solving until all constraints are known is that all constraints can be used for pruning the search space. So it might be the case that assuring global consistency at each step of the computation requires exponential computation time w.r.t. the problem size, whereas deciding consistency of the complete constraint set at the end is trivial. However, we feel that for a large fraction of practical problems this is not the case. In particular, this is true for optimization problems, which are formulated as

```
problem(<Variables>) :-
        <state constraints over Variables>,
        <maximize objective function over Variables>.
```

Typically, the constraint set that is generated by `<state constraints over Variables>` is easily shown to be consistent. Only optimizing the objective function is problematic. For these kind of problems, assuring global consistency is no problem.

Furthermore, we feel that if executing a CLP-program generates an unsatisfiable set of constraints, one should be immediately informed. This enables one to react at the precise point of failure. For example, consider the following fragment of a CLP-program.

```
problem(<Variables>) :-
        constraintA(<Variables>),
        constraintB(<Variables>),
        constraintC(<Variables>).

constraintB(<Variables>) :-
        <state constraints over Variables>.
constraintB(<Variables>) :-
        <react to failure>.
```

Given the common execution mechanism of logic programs, the programmer intends to react immediately in `<react to failure>` to an inconsistency encountered when adding the constraints of the first rule of `constraintB` to the current constraint set. Suppose that adding the constraint of the first rule of `constraintB` generates an unsatisfiable set of constraints, which is of course detected by a constraint solver assuring global consistency. As intended by the programmer, `<react to failure>` is invoked. Assume that a constraint solver based on local consistency techniques does not detect inconsistency, then computation goes on and the constraints of **constraintC** are added. Inconsistency is then detected by the extra enumeration predicate, but the precise point of failure is not known. Suppose that **constraintC** invoked heavy computation; it was useless, since it must be undone by backtracking in order to execute **<react to failure>**. In this case, providing a complete solver increases efficiency. Suppose that **constraintC** is implemented in the same way as **constraintB**, i.e., the programmer intends to react to failure caused by adding the constraints of **constraintC**. In case of inconsistency caused by **constraintB**, backtracking, caused by the failure of the extra enumeration predicate, will nevertheless first try to fix the problem by **<react to failure>** of **constraintC**, which is not what the programmer intended.

2.4.3 Solved Form and Entailment

For a CLP-program the simplified set of constraints collected at the end of execution is called the *answer constraint* and constitutes the solved form. It should provide a *good* and *explanative* description of all solutions of the current constraint set. An example is the answer constraint $A = 1$ to the previous query

$$?\text{-}A + B \geq 1, A + (1 - B) \geq 1.$$

In our opinion a good and explanative solved form should also allow entailment of constraints to be easily tested. Deciding entailment of constraints allows the interpretation of the presented results and enables the programmer to better understand the dependencies of the specified constraint problem. Furthermore, deciding entailment of constraints is essential for a new class of declarative programming languages; Concurrent Constraint Logic Programming (CCLP) languages [SHW93a, FHJ91], based on the ask and tell framework defined by Saraswat [Sar88, Sar89]. In CCLP-languages, deciding entailment and disentailment of constraints is needed for identifying determinate computation paths. Moreover, program execution can be guided not only by detecting inconsistency of a constraint set, but also by detecting entailment of constraints w.r.t. the current constraint set.

In CLP-languages based on local consistency techniques, like CHIP, there is no solved form for finite domain constraints and thus there is no obvious way to decide entailment of constraints, besides assuming the converse and trying to show unsatisfiability by enumeration. Providing a solved form is nevertheless necessary if we want to avoid an extra enumeration predicate. Furthermore, a solved form is typically more compact than an explicit enumeration of all possible solutions. For example, the query

$$?\text{-}\ A+B+C+D+E+F+G+H+I+J+K+L+M+ \\ N+O+P+Q+R+S+T+U+V+W+X+Y+Z \geq 13\ .$$

has $\sum_{i=13}^{26} \binom{26}{i} = 38,754,732$ different solutions, whereas the solved form is simply the constraint itself, naturally stating that at least 13 out of the 26 variables have to be 1. Constraints stating that at least n out of m variables have to be 1, appropriate for combinatorial problems, are the kind of constraints that are allowed in our solved form.

2.4.4 Conclusion

Constraint logic programming is intended to provide a declarative programming environment where subproblems over a specific domain are handled by a domain specific constraint solver. The semantic problems of current PROLOG implementations, caused by the standard PROLOG execution of logic programs, should, whenever possible, not be additionally merged with semantic problems caused by incomplete constraint solvers. Providing a complete constraint solver surely enhances the declarative reading of CLP-programs and encourages writing of declarative programs. Furthermore, the underlying constraint solving techniques can be enhanced without changing the program behavior. Hence, there is a need for a CLP-language solving combinatorial problems. Despite the fact that a complete constraint solver necessarily has to solve an NP-complete problem at each step, computation time need not always be exponential. Local consistency based methods also solve NP-complete problems, although the possible exponential blow up is bound to happen in the enumeration phase. Although both approaches essentially attack the same kind of problems, their application areas differ. On one hand, if the specified problem can be attacked by the underlying complete solver, there is no need to use local consistency based techniques and write a specific solver. If global consistency is needed because there are problem aspects that call for the control of execution by deciding NP-complete problems or if a description of the complete solution set is required, a CLP-system including a complete solver is the only possible choice. On the other hand, if local consistency based techniques have proven to be applicable to a class of constraint satisfaction problems and special heuristics, enumeration orderings and other tricks must be included, then writing a constraint solver in a logic programming system including local consistency techniques is the right choice. Hence, there is no a priori better approach.

2.5 LOGIC PROGRAMMING WITH PSEUDO-BOOLEAN CONSTRAINTS

The constraint logic programming language CLP($\mathcal{PB}$), introduced by Bockmayr [Boc91, Boc92b], aims to be a logic programming environment suitable for various applications where 0-1 or pseudo-Boolean problems naturally arise, which is the case for many AI-applications and operations research [BB93]. In this section we formally define the constraint domain $\mathcal{PB}$ of pseudo-Boolean constraints and instantiate the $\mathcal{X}$ in the CLP($\mathcal{X}$)-scheme to $\mathcal{PB}$. We then illustrate modelling combinatorial problems and programming techniques in CLP($\mathcal{PB}$) by some typical examples.

2.5.1 CLP($\mathcal{PB}$)

A pseudo-Boolean function is a mapping $f : \{0,1\}^n \to \mathbb{Z}$, where $\mathbb{Z}$ denotes the ring of integer numbers. Pseudo-Boolean functions are a generalization of Boolean functions, since the codomain of Boolean functions, $\{0,1\}$, is enlarged to $\mathbb{Z}$. First, we define a signature that allows the representation of pseudo-Boolean functions as pseudo-Boolean terms.

Definition 2.5.1 [Boc91] Let Σ_{psbool} be a signature with the sort symbol integer and the set of function symbols

$$F_{\mathsf{psbool}} := \begin{array}{rcl} \{ \quad 0,1 & : & \to \mathsf{integer}, \\ +,-,\cdot & : & \mathsf{integer} \times \mathsf{integer} \to \mathsf{integer}, \\ - & : & \mathsf{integer} \to \mathsf{integer}\} \end{array}$$

and let

$$\mathsf{Var}_{\mathsf{bool}} := \{X, Y, Z, X_1, X_2, \ldots\}$$

denote a countable infinite set of Boolean variables, i.e., the domain of the variables is $\{0,1\}$. The terms in the term algebra $\mathcal{T}(F_{\mathsf{psbool}}, \mathsf{Var}_{\mathsf{bool}})$ are called *pseudo-Boolean terms*. By interpreting the function symbols $+, \cdot, -$ as the usual arithmetic operations on integer numbers any pseudo-Boolean term $t \in \mathcal{T}(F_{\mathsf{psbool}}, \mathsf{Var}_{\mathsf{bool}})$ containing n Boolean variables *generates* an n-ary pseudo-Boolean function $f : \{0,1\}^n \to \mathbb{Z}$.

We allow the abbreviation $n \cdot t$ for $\overbrace{t + \cdots + t}^{n\times}$, where n is a positive integer and t is a pseudo-Boolean term. Next, we define constraint relations between pseudo-Boolean terms that are allowed in our constraint logic programming language.

Definition 2.5.2 We extend the signature Σ_{psbool} to Σ by adding the binary predicate symbols

$$P_{\mathsf{psbool}} := \{=, <, >, \leq, \geq : \mathsf{integer} \times \mathsf{integer}\} \ .$$

Predicates of the form $t_1 \bowtie t_2$, where t_1 and t_2 are pseudo-Boolean terms and $\bowtie \in P_{\mathsf{psbool}}$ are called *primitive constraints.* Let $\mathfrak{L}$ be the constraints generated by the primitive constraints and let $\mathfrak{D}$ interpret the symbols of Σ as usual in arithmetics. Then $\mathcal{PB} = (\mathfrak{D}, \mathfrak{L})$ is the pseudo-Boolean constraint domain.

The structure $\mathcal{PB}$ is solution compact, since it has no limit elements (limit elements are those that can be defined only by an infinite number of constraints [JL87]). From the general results of Jaffar and Lassez [JL87, JM94] for CLP($\mathcal{X}$) we obtain a semantic basis (see page 15) for the constraint logic programming language CLP($\mathcal{PB}$).

2.5.2 Programming in CLP($\mathcal{PB}$)

A typical CLP($\mathcal{PB}$)-program, namely **sequence**, has already been presented in Section 2.1. Another example, coming from operations research, is the following knapsack problem [Boc91]. Suppose there is a vessel with capacity w and goods g_i with weight w_i and value v_i for $i = 1, \ldots, n$. Let us introduce a Boolean variable X_i for all $1 \leq i \leq n$ that indicates whether the goods g_i are loaded on the vessel. The possible cargos not exceeding the capacity of the vessel can be determined by the following constraint rule.

$$\begin{aligned} &\mathbf{cargo}(X_1, \ldots, X_n) \text{ :-} \\ &\qquad w_1 \cdot X_1 + \cdots + w_n \cdot X_n \leq w. \end{aligned}$$

The goal $\mathbf{cargo}(X_1, \ldots, X_n)$ adds the constraint over the variables $X_1, \ldots, X_n$ that only admissible cargos are consistent. If we want to find the most valuable cargo, this can be done using a meta-predicate max /2 that maximizes a pseudo-Boolean function subject to some goal.

$\texttt{most\_valuable\_cargo}(X_1,\ldots,X_n)$:-
  $\max(v_1 \cdot X_1 + \cdots + v_n \cdot X_n, \texttt{cargo}(X_1,\ldots,X_n))$.

The semantic problems introduced by the meta-predicate max are discussed in [Fag93, MS94].

Next, we present a typical problem where a pseudo-Boolean formulation is suitable and show how to solve the problem with a corresponding CLP($\mathcal{PB}$)-program. The problem is to assemble a radio set under the following conditions [Boc91, HR68].

- Any one of the three types T_1, T_2, T_3 of tubes may be utilized, but only one.
  $\rightsquigarrow T_1 + T_2 + T_3 = 1$
- The box may be either of wood W, or of plastic material P.
  $\rightsquigarrow W + P = 1$
- When using P, dimensionality requirements impose the choice of T_2, and as there is no place for a transformer F, a special power supply S is needed.
  $\rightsquigarrow P \leq T_2 \cdot S$
- T_1 needs F.
  $\rightsquigarrow T_1 \leq F$
- when choosing T_2 or T_3 we need S
  $\rightsquigarrow T_1 + T_2 \leq S$
  (and not F).
  $\rightsquigarrow F + S = 1$

The prices of the above components are:

one series of tubes T_1	28 units
one series of tubes T_2	30 units
one series of tubes T_3	31 units
one transformer F	25 units
one special power supply S	23 units
one wooden box W	9 units
one plastic material box P	6 units

The other necessary components of the radio set cost

27 units,	if the tubes T_1 are utilized,
28 units,	if the tubes T_2 are utilized,
25 units,	if the tubes T_3 are utilized.

The price of production is 10 units for each set in all the cases and a set is sold at 110 units when it is enclosed in a wooden box, and at 105 units in the other case. Which model is to be constructed in order to maximize the profit? The following CLP-program implements the specification.

```
assemble(T1, T2, T3, F, S, W, P) :-
        T1 + T2 + T3 = 1,
        W + P = 1,
        P ≤ T2 · S,
        T1 ≤ F,
        T1 + T2 ≥ S,
        F + S = 1.

maximal_profit(T1, T2, T3, F, S, W, P) :-
        max(110 · W + 105 · P−
                (28 · T1 + 30 · T2 + 31 · T3 + 25 · F + 23 · S + 9 · W+
                 6 · P + 27 · T1 + 28 · T2 + 25 · T3 + 10),
                      assemble(T1, T2, T3, F, S, W, P).
```

The query

```
?-maximal_profit(T1, T2, T3, F, S, W, P).
```

yields the unique answer

$$T_1 = 0, T_2 = 0, T_3 = 1, F = 0, S = 1, W = 1, P = 0.$$

Interesting perspectives of the specified problem can be obtained by fixing some components of the problem and investigating the generated solved form. For example, the query

```
?-assemble(T1, 1, T3, F, S, W, P).
```

gives rise to the answer

$$T_1 = 0, T_3 = 0, F = 0, S = 1, W + P = 1.$$

from which we conclude that using tubes T_2 left no choices, except that of the material of the box.

Insight about a specified problem can be obtained by interpreting the declarative answer constraint returned by the CLP-system. For example, suppose that we want to install an alarm system containing 10 equivalent detectors [BB93]. Each of them has two possible states `OK` or `ALERT`. The problem is to deduce a warning level that can be either `green`, `orange` or `red`. The specification defines the warning level `red` if the state of 5 or more detectors is `ALERT`, `orange` if the number of detectors that are in the state `ALERT` lies between 1 and 4 and `green` if no detector is in the state `ALERT`. With each of the detectors we associate a Boolean variable D_i, which has the value 0 if the detector is in the state `OK` and the value 1 if the detector is in the state `ALERT`. This specification can be implemented in CLP($\mathcal{PB}$) by the following rules

$$\begin{array}{l}
\texttt{warning_level}(D_1, D_2, \ldots, D_9, D_{10}, \texttt{red}) :- \\
\qquad D_1 + D_2 + \cdots + D_9 + D_{10} \geq 5. \\
\texttt{warning_level}(D_1, D_2, \ldots, D_9, D_{10}, \texttt{orange}) :- \\
\qquad D_1 + D_2 + \cdots + D_9 + D_{10} \leq 4, \\
\qquad D_1 + D_2 + \cdots + D_9 + D_{10} \geq 1. \\
\texttt{warning_level}(D_1, D_2, \ldots, D_9, D_{10}, \texttt{green}) :- \\
\qquad D_1 + D_2 + \cdots + D_9 + D_{10} = 0.
\end{array}$$

To compute the warning level we use a logical variable `Warninglevel`, which is instantiated by the query

$$\texttt{?-warning\_level}(D_1, \ldots, D_{10}, \texttt{Warninglevel}).$$

where $D_i \in \{0, 1\}$ for all $1 \leq i \leq 10$. Suppose now that the detectors associated with D_2, D_5, and D_9 are broken and hence no value is assigned. Furthermore, suppose that D_1, D_6, and D_7 are in the state `ALERT` and the remaining ones are in the state `OK`. Then the query

$$\texttt{?-warning\_level}(1, D_2, 0, 0, D_5, 1, 1, 0, D_9, 0, \texttt{Warninglevel}).$$

can succeed with the following two answer constraints

$$D_2 + D_5 + D_9 \geq 2, \texttt{Warninglevel} = \texttt{red}$$
$$\overline{D_2} + \overline{D_5} + \overline{D_9} \geq 2, \texttt{Warninglevel} = \texttt{orange}$$

So unifying `Warninglevel` with warning level `red` succeeds, which is important for safety-critical applications, where detectors may indeed fail. Furthermore, we derive from the answer constraints that if at least two of the three failed detectors are in the state `ALERT`, we have warning level `red` and on the other hand, that if at least two of the three failed detectors are in the state `OK`, we ensure warning level `orange`.

Suppose that D_2, D_5, and D_9 are repaired and that all are in the state `OK`. This can be modelled by adding the constraints $D_2 = 0, D_5 = 0$, and $D_9 = 0$. Since then the constraint $D_2 + D_5 + D_9 \geq 2$ is violated, the system correctly deduces the only possible answer constraint `Warninglevel = orange`.

As demonstrated in this `warning_level`-example, using classical Boolean constraints does not always yield the most adequate problem description. For example, constraints like "at least or at most d out of n facts need to hold" cannot be expressed directly as Boolean constraints. On the other hand, representing classical Boolean constraints as pseudo-Boolean constraints is straightforward. Thus, pseudo-Boolean constraints often allow a given problem to be more naturally expressed.

For example, suppose that D_1, D_2, D_3, D_4, and D_5 represent detectors as in the `warning_level`-example and that the following Boolean specification defines the warning level `red`.

$$(D_1 \vee (D_2 \wedge (D_3 \vee D_4)) \vee (D_3 \wedge D_4 \wedge d_5)) \quad \Longrightarrow \quad \texttt{red} \tag{2.1}$$

We can immediately translate this specification into an equivalent pseudo-Boolean specification, namely

$$D_1 + D_2 \cdot (D_3 + D_4) + D_3 \cdot D_4 \cdot D_5 \geq 1.$$

If the number of detectors and the number of conditions increase, then Boolean specifications become more and more complicated and difficult, or even impossible, to understand due to their size. In that case a more compact and understandable specification may become necessary. For example, we can introduce weights for each detector expressing its importance. Deciding the warning level is then done by giving lower bounds for some combinations of detectors. In the example above, the pseudo-Boolean constraint

$$5 \cdot D_1 + 3 \cdot D_2 + 2 \cdot D_3 + 2 \cdot D_4 + D_5 \geq 5$$

is again equivalent to the Boolean specification (2.1). A complete specification can be investigated by studying the solved form, i.e., checking the entailed constraints.

Finally, we present some generic examples that can be used in several applications, which show how to implement in CLP($\mathcal{PB}$) some typical problems arising in the modelling phase of a combinatorial problem [Wil90]. In many applications one wants to express a logical relationship between Boolean variables like conjunction, disjunction, etc.. In the standard reference on modelling [Wil90], we can find how to express these logical conditions using linear pseudo-Boolean constraints.

$$
\begin{array}{ll}
\texttt{or}(X_1, X_2) \text{ :- } & X_1 + X_2 \geq 1. \\
\texttt{and}(X_1, X_2) \text{ :- } & X_1 + X_2 = 2. \\
\texttt{implies}(X_1, X_2) \text{ :- } & X_1 - X_2 \leq 1. \\
\texttt{equiv}(X_1, X_2) \text{ :- } & X_1 - X_2 = 1.
\end{array}
$$

With these declarations we can now use an appropriate name whenever we intend to use a logical connective and we are not forced to translate it directly to a linear pseudo-Boolean constraint. This increases readability. Note that since CLP($\mathcal{PB}$) also accepts non-linear pseudo-Boolean constraints, a thorough treatment of the conversion of complex logic relations into linear pseudo-Boolean constraints [MW89] is not necessary, although possible.

Another common modelling problem is that we want to decide with the help of a Boolean variable, whether a linear pseudo-Boolean inequality should hold or whether it may be violated. Again in [Wil90], we find that for a linear pseudo-Boolean inequality

$$L \geq R \tag{2.2}$$

the linear pseudo-Boolean inequality

$$-M \cdot \overline{B} + L \geq R \ , \tag{2.3}$$

where M is a lower bound of $L - R$ and B is a new Boolean variable, is the desired formulation. If B is 1 then (2.3) is equivalent to (2.2). If B is 0 then (2.3) is equivalent to

$$L - R \geq M,$$

which always holds, since M is a lower bound of $L - R$. Hence, setting B to 1 activates the constraint and setting B to 0 deactivates it. We give a CLP($\mathcal{PB}$)-program that attaches a Boolean variable B to a linear pseudo-Boolean inequality with the above described behavior.

```
switchconstraint(L ≥ R, B) :-
        lowerbound(L − R, M),
        −M · B̄ + L ≥ R.
```

A simple implementation of `lowerbound` computes the sum of all negative coefficients of the linear pseudo-Boolean term. Having `switchconstraint` available we can express a disjunction between two constraints C_1 and C_2. We model this by introducing two switching variables B_1 and B_2 for C_1 and C_2 respectively. Between these switching variables, the logical or relation is established.

```
orcs(C1, C2) :-
        switchconstraint(C1, B1),
        switchconstraint(C2, B2),
        or(B1, B2).
```

Since we have a complete constraint solver providing a nice solved form, we can also find out to what the disjunction of two constraints is equivalent. For example, the query

$$\texttt{orcs}(4 \cdot A + 3 \cdot B + 2 \cdot C \geq 3, 4 \cdot A + 2 \cdot B + 3 \cdot C \geq 3).$$

yields the answer

$$A + B + C \geq 1 \ \texttt{yes}.$$

Hence, a possible use of CLP($\mathcal{PB}$) is the preprocessing of disjunctive constraints in order to avoid the introduction of additional Boolean variables. A more general tool in modelling are cardinality constraints, which state that from a list of constraints at least L constraints and at most U constraints must hold. Using `switchconstraint`, implementing a generic cardinality constraint is straightforward.

```
card(L,Cs,U) :-
        switchsum(Cs,S),L ≤ S, S ≤ U.

switchsum([],0).
switchsum([C | Cs],S + B) :-
        switchconstraint(C, B), switchsum(Cs,S).
```

Let Cs be a list $[C_1, \ldots, C_n]$ of pseudo-Boolean constraints. The predicate `switchsum` associates a Boolean variable B_i to each constraint, with help of `switchconstraint`. In the second argument S of `switchsum` the pseudo-Boolean term $\sum_{i=1}^{n} B_i$ is constructed. In order to ensure that at least L and at most U constraints must hold we add the constraints $L \leq S$ and $S \leq U$ respectively. We can now model a disjunction of two constraints by

orcs(C_1, C_2) :-
    card$(1,[C_1, C_2],2)$.

Of course a solved form can be inspected, which may help to understand and improve the model. For example, we obtain for the disjunction

?- **card**$(1,[\, 4 \cdot A + 3 \cdot B + 2 \cdot C \geq 3, 4 \cdot A + 2 \cdot B + 3 \cdot C \geq 3 \,],2)$.
$A + B + C \geq 1$ **yes**.

and for the conjunction

?- **card**$(2,[\, 4 \cdot A + 3 \cdot B + 2 \cdot C \geq 3, 4 \cdot A + 2 \cdot B + 3 \cdot C \geq 3 \,],2)$.
$A + B \geq 1, B + C \geq 1$ **yes**.

The flexible and powerful modelling technique using cardinality is made generically available with just a few lines of code, which shows the merits of the combination of constraint solving and logic programming.

2.6 EXISTING CLP-SYSTEMS FOR COMBINATORIAL OPTIMIZATION

Combinatorial problems arise in a large number of applications. They appear in such diverse areas as in operations research (scheduling and assignment), hardware design (circuit verification and layout), decision making (portfolio management and option trading), or even biology (DNA-sequencing). Consequently, there is not a single general method for solving all kinds of combinatorial problems, but a wide range of specific problem classes together with dedicated constraint solving techniques. This diversity is also reflected in the different instances of the CLP-scheme, where each instance has its advantages and disadvantages w.r.t. the problem to be solved. We organize this overview of CLP-systems by constraint domain. For each constraint domain typical application areas and CLP-systems that support this constraint domain are mentioned.

First, we introduce finite domain constraint solving in logic programming based on local consistency techniques. Most of the successful, and in the meantime also commercial, applications have been developed with such systems. For Boolean problems typically a complete constraint solver is used, which has its main application in hardware verification. Although the whole book is about pseudo-Boolean constraints, we also list pseudo-Boolean constraints in this summary but with other constraint solving techniques proposed. Linear arithmetic constraints are typically completely solved inside CLP, but can also be used for combinatorial problems by viewing the global consistency w.r.t. the linear arithmetic constraint domain as relaxation w.r.t. the integer constraint domain and relying on enumeration. Finally, we briefly mention some other constraint domains.

2.6.1 Finite Domain Constraints

We illustrate solving finite domain problems with help of local consistency in a logic programming language on a well-known crypto-arithmetic problem; the "SEND+MORE=MONEY" example [Col86, vH89b]. The problem is to assign to each digit in the set {S,E,N,D,M,O,R,Y} a different digit from {0,...,9} such that the equation

$$\begin{array}{cccccc} & & S & E & N & D \\ + & & M & O & R & E \\ = & M & O & N & E & Y \end{array}$$

is satisfied. A logic program for solving this crypto-arithmetic problem consists of the two parts constraint generation and enumeration.

```
sendmory([S,E,N,D,M,O,R,Y]) :-
        generate_constraints([S,E,N,D,M,O,R,Y],[C1,C2,C3,C4]),
        enumerate([C1,C2,C3,C4,S,E,N,D,M,O,R,Y]).
```

First, we restrict the domain of the variables in the set {S,E,N,D,M,O,R,Y} to {0,...,9}. We also need carries in order to add two digits. The carries {C1,C2,C3,C4} are hence restricted to the domain {0,1}. Next, we state that all digits must be pairwise different and finally the arithmetic conditions between the digits and the carries are written down.

```
generate_constraints([S,E,N,D,M,O,R,Y],[C1,C2,C3,C4]) :-
        [S,E,N,D,M,O,R,Y] :: [0 .. 9],
        [C1,C2,C3,C4] :: [0 .. 1],
        alldiff([S,E,N,D,M,O,R,Y],
        S ≠ 0,
        M ≠ 0,
                  C1 = M ,
         C2 + S + M = O + 10·C1,
         C3 + E + O = N + 10·C2,
         C4 + N + R = E + 10·C3,
              D + E = Y + 10·C4.
```

Enumeration is implemented by

```
enumerate([V|Vs]) :-
        indomain(V),
        enumerate(Vs).
```

where **indomain** is typically a built-in predicate that assigns a value from the current domain of V to V. Via backtracking all possible values in the current domain are tried. The query

```
?- sendmory([S,E,N,D,M,O,R,Y]).
```

then yields the, in this case unique, answer

$$S = 9, E = 5, N = 6, D = 7, M = 1, O = 0, R = 8, Y = 2 \quad .$$

Given a complete constraint solver for finite domain constraints the enumeration part could have been skipped. The local consistency solver only eliminates impossible values from the domain of the variables which are detected by some consistency procedure. For a detailed introduction into consistency techniques we refer to [vH89a, Kum92, Tsa93]. We follow the first steps of a typical propagation mechanism. Since $M \neq 0$, $C1 = M$, and C1 is either 0 or 1, we derive that

M = C1 = 1. The value 1 is removed from all the variables in {S,E,N,D,O,R,Y} due to the `alldiff` constraint. Since M = C1 = 1, the constraint

$$C2 + S + M = O + 10 \cdot C1$$

reduces to

$$C2 + S = O + 9 \quad .$$

Since O + 9 is greater than or equal to 9 in all possible assignments, we conclude that S must be greater than or equal to 8, because C2 is at most 1. Hence, the domain of S is reduced to {8,9}. Furthermore, the domain of O is reduced to {0}, since C2 + S is at most 10. Hence, O must be smaller than 2, but 1 is already eliminated from the domain. Proceeding further we obtain the following state before starting the enumeration.

$$S{=}9,\ M{=}1,\ O{=}0,\ C1{=}1,\ C2{=}0, \text{ and } E,N,D,R,Y \in \{2,\ldots,8\}$$

While enumerating we choose a value from a not yet fully determined finite domain variable and fix this variable. With this information the domains of the other variables are further reduced until either a solution is found or inconsistency is detected in which case backtracking takes place. What is still missing is an implementation of `alldiff`.

```
alldiff([]).
alldiff([X|Y]) :-
        outof(X,Y),
        alldiff(Y).

outof(X,[]).
outof(X,[F|T]) :-
        X ≠ F,
        outof(X,T).
```

The predicate `alldiff` imposes the constraint X ≠ Y for all pairs of elements in the list in its argument. Note that we are creating a quadratic number of constraints w.r.t. the number of variables. But, the `alldiff` constraint could also be handled algorithmically. Whenever a value is assigned to a variable in an `alldiff` relation, then for all remaining variables this value is removed. Such an algorithmic treatment avoids the creation of a quadratic number of constraints while the same pruning effect is achieved. Furthermore, even better pruning can be obtained by using information about the `alldiff` relation. The property that for every subset of k variables the cardinality of the union of the associated domains must be greater than or equal to k [Hal35] can be exploited, which is done in commercial systems like CHIP [DvHS+88a] and ILOG-Solver [Pug94].

There are other common complex constraints, so-called *global constraints*[6], which occur very often in practical applications. For these constraints often an effective algorithmic treatment is possible [BC94, AB93, BLPN95]. Among them are the `cumulative` constraint, which handles typical resource constraints occurring in scheduling problems, the `among` constraint, which is a generalization of the `atleast` and `atmost` constraints stating that from a list of constraints at least or at most n constraints must hold, the `diffn` constraint, a generalization of the `alldiff` constraint, and the `cycle` constraint, which is designed for solving complex vehicle routing problems. There are several advantages of introducing these special purpose constraints into a logic programming language:

- Special purpose algorithms for specific problem types coming from operations research are made available. Hence, efficiency is greatly enhanced while declarativity is still preserved.
- Due to their integration in the propagation paradigm these special purpose algorithms can be combined. Hence, complex problems where several special constraint types occur can be handled.

The propagation-based approach for solving combinatorial problems requires a good knowledge of the problem to be solved and of the techniques used inside the constraint solver in order to attack the problem with the right propagators.

Application areas of finite domain constraint solving with local consistency techniques inside a logic programming language include car sequencing [vH89a, DSvH88c], scheduling [DSvH88b, vH92, Wal94, LP94, BC94, BL95, BLPN95], cutting stock problems [DSvH88a, vH89a], test pattern generation [vHSD92], graph coloring [vH89a], microcode label assignment [vH89a], financial decision

[6] Although such a special treatment of constraints typically results in a better pruning and hence reduces the search space, the word *global* has nothing to do with ensuring global consistency.

making [Ber90], planning [Ber88, BDP93], frequency assignment for cellular telephones [CG93], and many others.

Finite domain constraint handling techniques are integrated in many CLP-systems including CHIP [DvHS+88a], ECLIPSE [MAC+89], ECHIDNA [HJO+92], CLP($\mathcal{FD}$) [DC93], IFPROLOG [IFP95], and OZ [SSW94]. SICStus [CW90] should also have finite domain constraints included at the time you hold this book in your hands. There are also systems that emphasize on the constraint programming aspect and no longer include a logic programming language like ILOG [Pug94] and CHARME [OMT89].

2.6.2 Boolean Constraints

Büttner and Simonis [BS87] embedded Boolean constraints into logic programming. Many constraint solvers for Boolean problems are based on Boolean unification [BS87, MN88, MN89] (cf. Section 3.2.2) and are implemented with help of binary decision diagrams [Bry86, Bry92]. They provide complete constraint solving and a solved form. The constraint solver handles equations between Boolean terms over Boolean variables and the connectives $\vee, \wedge$, and $\oplus$ as in Definition 3.1.5 For example, we show how to model a digital circuit, here, a full adder, in a constraint logic programming language over Boolean constraints [BS87].

We describe Boolean circuits by predicates whose arguments represent the Boolean signals. The logic programming language then allows one to build complex Boolean circuits by using simpler ones. As primitive circuits we have the built-in Boolean connectives. First, we implement a half adder.

```
half_adder(X,Y,X ⊕ Y,X ∧ Y).
```

where X and Y represent the inputs. The third argument represents the sum, whose value is the **exor** of the input arguments. The fourth argument represents the carry, whose value is the **and** of the input arguments. A full adder can be built from two half adders where the connections of the input and output arguments are as follows.

```
full_adder(X,Y,Cin,S,C1 ∨ C2) :-
        half_adder(X,Y,S1,C1),
        half_adder(S1,Cin,S,C2).
```

The query

$$?\text{-}\mathtt{full\_adder}(X,Y,Cin,S,Cout).$$

yields a symbolic solution, i.e., a most general Boolean unifier, of the generated Boolean equations.

$$S = X \oplus Y \oplus Cin,$$
$$Cout = X \wedge Y \oplus X \wedge Cin \oplus Y \wedge Cin \ \mathtt{yes}.$$

We see that `full_adder` has the expected functionality. Queries like

$$?\text{-}\mathtt{full\_adder}(X,Y,Cin,1,0).$$
$$X = A$$
$$Y = A \wedge B \oplus B$$
$$C = A \wedge B \oplus A \oplus B \oplus 1 \ \mathtt{yes}.$$

show the potential of using Boolean unification embedded in logic programming for testing and analyzing digital circuits [SND88].

Boolean constraint solving based on Boolean unification is available in the logic programming systems CHIP [DvHS+88a] and SICStus [CW90]. The main application area is hardware verification, but whenever a complete symbolic description of all solutions of a Boolean problem is needed it might be applicable. Other methods for solving Boolean constraints are integrated in PROLOG III [Col90, Ben93] based on resolution and in CAL [ASS+88] based on Buchberger's algorithm for Gröbner bases. Boolean problems can also be handled by finite domain techniques if just one solution is needed and the problem class is sensible to finite domain techniques [SD90, CD93, CD94].

2.6.3 Pseudo-Boolean Constraints

Beside the logic-based approach for handling pseudo-Boolean constraints we can also exploit a polyhedral approach for pseudo-Boolean constraint solving inside CLP, as proposed by Bockmayr [Boc93c, Boc94] (see also Section 3.2.4). Furthermore, a generalization of Boolean unification to the pseudo-Boolean case [Boc93b] can be used (see also Section 3.2.2). Another possibility is to convert pseudo-Boolean constraints into Boolean constraints, which seems to be impractical because of the size of the generated constraints. Moreover, finite domain techniques based on local consistency techniques can also handle pseudo-Boolean constraints, which is discussed in Section 2.4 and Section 3.2.3.

2.6.4 Linear arithmetic

CLP($\mathcal{R}$) was one of the first instances of a CLP-system. Jaffar and Lassez [JL87] illustrated the CLP($\mathcal{X}$)-scheme by instantiating the computational domain $\mathcal{X}$ to linear arithmetic constraints. First, a CLP($\mathcal{R}$)-interpreter has been made available and later a compiler was developed [JMSY90, HJM+92], which also allowed meta-programming with linear arithmetic constraints [HMSY89]. The CLP($\mathcal{R}$)-system has been used for many problems, including electrical engineering problems [HMS87], options trading [LMY87, HL88], and temporal reasoning [Brz91].

Currently, many logic programming environments provide linear arithmetic constraint solving, among them CHIP [DvHS+88a], ECLIPSE [MAC+89], and SICStus [CW90, Hol95]. Many use rational constraints with infinite precision arithmetic in order to avoid numerical problems. All systems can also handle non-linear constraints by delaying them and reactivate them in case they become linear.

The underlying constraint solvers are specialized simplex algorithms combined with Gaussian elimination and Fourier-Motzkin elimination [JMSY90]. They provide complete constraint solving for linear arithmetic inequalities, equations, and disequations. We show the features of constraint logic programming with linear arithmetic constraints on the standard mortgage-example [HJM+92].

mortgage(P, Time, IntRate, Bal, MP) :-
    Time = 1,
    Bal = P $\cdot$ (1 + Time $\cdot$ $\frac{\text{IntRate}}{1200}$) - Time $\cdot$ MP.
mortgage(P, Time, IntRate, Bal, MP) :-
    Time > 1,
    mortgage(P $\cdot$ (1 + $\frac{\text{IntRate}}{1200}$) - MP, Time-1, IntRate, Bal, MP).

The parameters of a mortgage, i.e., the principal P, the duration of the mortgage in months Time, the annual interest rate in % IntRate, which is compounded monthly, the monthly payment MP, and the outstanding balance Bal, is related naturally by these two rules. In order to find out how much it would cost to finance a mortgage with a principal of 100000 at 12 percent for 15 years we ask the query

```
?-mortgage(100000, 180, 12, 0, MP).
```

which yields the answer

```
MP = 1200.17 yes.
```

The advantages of using CLP are again the bidirectionality and the output of constraints as answers. For example, the query

```
?-mortgage(P, 180, 12, Bal, MP).
```

yields the answer

```
P = 0.166783 * Bal + 83.3217 * MP yes.
```

naturally relating the principal to the balance and the monthly payment if the duration is 180 months and the interest rate is 12 %.

Combinatorial problems can be handled within CLP($\mathcal{R}$) by viewing the complete solver over linear arithmetic constraints as an approximative solver for integer domain constraints. Just ensuring local consistency is replaced by solving the linear programming relaxation of the integer programming problem. For example, the crypto-arithmetic problem in Section 2.6.1 runs with almost no changes[7]. CLP($\mathcal{R}$) should be preferred if solving the linear programming relaxation of a combinatorial problem turns out to achieve a better pruning than using local consistency techniques. The advantage of CLP($\mathcal{R}$) is that also mixed-integer programming problems can be handled. But, like in local consistency based systems, constraint solving relies on enumeration. Unfortunately, commercial linear programming systems are typically much faster than the constraint solvers inside a CLP-system for linear arithmetic [Cra93, CFGG95]. Hence, main application areas of CLP($\mathcal{R}$) are problems that either rely on the

[7] Just the domain constraints X :: [L .. U] have to be replaced by $X \geq L$ and $X \leq U$, and the indomain predicate has to be implemented.

declarative and simplified answer constraint or problems where exploiting the programming language aspect is necessary.

2.6.5 Other constraint domains

Several other constraint domains have been proposed that are more or less well suited for handling combinatorial problems.

Interval Constraints

Older [OV90] has introduced interval constraints into logic programming. Handling arithmetic constraints with interval techniques is based on the same idea as handling finite domain constraints with local consistency techniques, i.e., bound propagation combined with enumeration. The advantage of using interval constraints is that also non-linear arithmetic can be handled. Furthermore, interval arithmetic can also be used to solve discrete problems [BO93]. Interval constraints are available in CLP(BNR) [OB93], NEWTON [BMvH94], CIAL [CL94], and PROLOG-IV.

Set Constraints

Constraints over finite sets are available as an extension of ECLIPSE [MAC+89, Ger94]. Their application areas are combinatorial problems based on sets, relations, or graphs like warehouse location, resource allocation, bin packing and others [Ger94]. They may provide a more natural formulation of a problem than a finite domain or (0-1) integer programming formulation.

Non-linear Arithmetic Constraints

As already mentioned interval methods can also handle non-linear constraints. CLP-systems that support linear arithmetic constraints accept also non-linear constraints, but these are typically completely delayed and not solved. However, there is one CLP-system, RISC-CLP($\mathcal{R}$) [Hon92], that is able to handle non-linear constraints over the real numbers with computer algebra techniques based on quantifier elimination (Partial Cylindrical Algebraic Decomposition) [Tar51, Col75, Hon90]. The applicability of these constraint solving techniques to *just* NP-hard problems seems limited[8].

[8]Hoon Hong once said: "Other people build black-box solvers, we build black-hole solvers. Once called they never return."

3

PSEUDO-BOOLEAN CONSTRAINTS

After a formal presentation of the constraint domain $\mathcal{PB}$ we investigate several methods for solving pseudo-Boolean constraints and examine their applicability in the context of constraint logic programming. We adopt a logical view of pseudo-Boolean constraints and propose a solved form suitable for the constraint logic programming language CLP($\mathcal{PB}$).

3.1 PRELIMINARIES

3.1.1 Basic Definitions

Pseudo-Boolean Terms

With the signature Σ_{psbool} of Definition 2.5.1 we can represent pseudo-Boolean functions as pseudo-Boolean terms. A unique representation of a pseudo-Boolean term is the *polynomial normal form* of pseudo-Boolean functions.

Proposition 3.1.1 *[HR68] For any pseudo-Boolean function $f : \{0,1\}^n \rightarrow \mathbb{Z}$ there is a unique, up to associativity and commutativity of $+$ and $\cdot$, pseudo-Boolean term of the form*

$$\sum_{I \in \Omega} c_I \prod_{i \in I} X_i \ , \tag{3.1}$$

where Ω is a collection of subsets of $\{1, \ldots, n\}$ and the c_I are coefficients in $\mathbb{Z} \setminus \{0\}$, that generates f.

We shall adopt the usual convention of identifying a pseudo-Boolean term and the pseudo-Boolean function it generates.

Example 3.1.2 Let the function $f : \{0,1\}^3 \to \mathbb{Z}$ be defined by the following table.

X	0	0	0	0	1	1	1	1
Y	0	0	1	1	0	0	1	1
Z	0	1	0	1	0	1	0	1
f(X,Y,Z)	-5	0	0	0	-3	8	0	6

The polynomial normal form of $f(X, Y, Z)$ is

$$-5 + 2 \cdot X + 5 \cdot Y + 5 \cdot Z - 2 \cdot X \cdot Y + 6 \cdot X \cdot Z - 5 \cdot Y \cdot Z \ .$$

An important class of pseudo-Boolean terms are linear pseudo-Boolean terms.

Definition 3.1.3 Let $\sum_{I \in \Omega} c_I \prod_{i \in I} X_i$ be the polynomial normal form of a pseudo-Boolean term t. The pseudo-Boolean term t is *linear* if $|I| \leq 1$ for all $I \in \Omega$.

We introduce the abbreviation $\overline{X_i}$ for the pseudo-Boolean term $(1 - X_i)$, where X_i is a Boolean variable. We say $\overline{X_i}$ is the *negation* of the Boolean variable X_i. A *literal* L_j is either a Boolean variable X_i or the negation of a Boolean variable $\overline{X_i}$. Let $\mathcal{L}$ be the set of all literals. The negation of a negative literal $\overline{\overline{X_i}}$ is always simplified to X_i. We denote by $\mathsf{Var}(X_i) = \mathsf{Var}(\overline{X_i}) = X_i$ the variable of a literal.

Definition 3.1.4 An *assignment* is a mapping $\alpha : \mathsf{Var}_{\mathsf{bool}} \to \{0,1\}$. We extend an assignment to a mapping $\alpha : \mathcal{T}(F_{\mathsf{psbool}}, \mathsf{Var}_{\mathsf{bool}}) \to \mathbb{Z}$ by defining

$$\begin{array}{rcl} \alpha(0) & := & 0 \\ \alpha(1) & := & 1 \\ \alpha(t_1 + t_2) & := & \alpha(t_1) + \alpha(t_2) \\ \alpha(t_1 \cdot t_2) & := & \alpha(t_1) \cdot \alpha(t_2) \\ \alpha(t_1 - t_2) & := & \alpha(t_1) - \alpha(t_2) \\ \alpha(-t_1) & := & -\alpha(t_1) \ , \end{array} \tag{3.2}$$

where $t_1, t_2 \in \mathcal{T}(F_{\mathsf{psbool}}, \mathsf{Var}_{\mathsf{bool}})$ and $+, -$, and $\cdot$ have the usual meaning of addition, subtraction, and multiplication respectively in $\mathbb{Z}$.

Boolean functions, i.e., mappings $g : \{0,1\}^n \to \{0,1\}$, are a restriction of pseudo-Boolean functions. We introduce a signature with the usual Boolean operators that allows the representation of Boolean functions as Boolean terms.

Definition 3.1.5 [Boc91] Let Σ_{bool} be a signature with the sort symbol $\mathsf{bool} \subseteq \mathsf{integer}$ and the set of function symbols

$$F_{\mathsf{bool}} := \begin{array}{rcl} \{ \quad 0,1 & : & \to \mathsf{bool}, \\ \bar{\ } & : & \mathsf{bool} \to \mathsf{bool}, \\ \wedge, \vee, \oplus & : & \mathsf{bool} \times \mathsf{bool} \to \mathsf{bool} \} \ . \end{array}$$

The terms in the term algebra $\mathcal{T}(F_{\mathsf{bool}}, \mathsf{Var}_{\mathsf{bool}})$ are called *Boolean terms.* By interpreting $\bar{\ }$ as negation, $\oplus$ as sum modulo 2, $\wedge$ as conjunction, and $\vee$ as disjunction, any pseudo-Boolean term $t \in \mathcal{T}(F_{\mathsf{bool}}, \mathsf{Var}_{\mathsf{bool}})$ containing n Boolean variables *generates* an n-ary Boolean function $f : \{0,1\}^n \to \{0,1\}$. Again, we adopt the convention to identify a Boolean term and the Boolean function it generates.

We can represent every Boolean term t as an equivalent pseudo-Boolean term by normalizing t with the following canonical term rewrite system.

$$\begin{array}{rcl} X \oplus Y & \to & X + Y - (X \cdot Y + X \cdot Y) \\ X \wedge Y & \to & X \cdot Y \\ \overline{X} & \to & 1 - X \\ X \vee Y & \to & X + Y - X \cdot Y \end{array}$$

We denote by $PB(t)$ the pseudo-Boolean term obtained by applying the above given rewrite system on a Boolean term t.

Pseudo-Boolean constraints

Pseudo-Boolean constraints are equations or inequalities between two pseudo-Boolean functions. They are represented as pseudo-Boolean constraints between two pseudo-Boolean terms according to Definition 2.5.2.

Definition 3.1.6 A pseudo-Boolean constraint $t_1 \bowtie t_2$ with $\bowtie \in P_{\mathsf{psbool}}$ and $t_1, t_2 \in \mathcal{T}(F_{\mathsf{psbool}}, \mathsf{Var}_{\mathsf{bool}})$ is *satisfied* by an assignment α if

$$\alpha(t_1) \bowtie \alpha(t_2)$$

holds, where $\bowtie \in P_{\mathsf{psbool}} = \{=, <, >, \leq, \geq : \mathsf{integer} \times \mathsf{integer}\}$ has its usual meaning in $\mathbb{Z}$. An assignment satisfying a pseudo-Boolean constraint is also called a *solution* of the pseudo-Boolean constraint. A pseudo-Boolean constraint $t_1 \bowtie t_2$ is *satisfiable* if there exists a solution α of $t_1 \bowtie t_2$.

A pseudo-Boolean constraint that is satisfied by any assignment is called *tautology* and simplified to $\top$. A pseudo-Boolean constraint that is not satisfiable is called *contradiction* and simplified to $\bot$.

The *extension* $\mathsf{Ext}(C)$ of a pseudo-Boolean constraint C is the set of assignments satisfying C. The extension of a set of pseudo-Boolean constraints S is the intersection of the extensions of the pseudo-Boolean constraints in S. A set of pseudo-Boolean constraints is satisfiable if $\mathsf{Ext}(S)$ is non-empty. A set of pseudo-Boolean constraints S *(strictly) dominates* a set of pseudo-Boolean constraints S' if $\mathsf{Ext}(S)$ is a (proper) subset of $\mathsf{Ext}(S')$. We also write $S \models S'$ if S dominates S'. We abbreviate $\{C\} \models S'$ by $C \models S'$ and $S \models \{C\}$ by $S \models C$. Two sets of pseudo-Boolean constraints S, S' are *equivalent* if $\mathsf{Ext}(S) = \mathsf{Ext}(S')$. If a set of pseudo-Boolean constraints S dominates a pseudo-Boolean constraint C, we say C is *valid* w.r.t. S.

Solving a set of pseudo-Boolean constraints means to check it for satisfiability and, optionally, to represent the set of all its solutions, i.e., its extension.

Optimization

In practical applications an important problem is to solve a *pseudo-Boolean program*, which means minimizing or maximizing a pseudo-Boolean function, possibly subject to a set of pseudo-Boolean constraints.

Definition 3.1.7 Let C be a set of pseudo-Boolean constraints and let t be a pseudo-Boolean term. An assignment $\alpha \in \mathsf{Ext}(C)$ is *globally maximizing* for C if

$$\alpha(t) \geq \alpha'(t) \quad \text{for all } \alpha' \in \mathsf{Ext}(C) \ .$$

The value $\alpha(t)$ for a globally maximizing α is called the *maximum* of t subject to C.

Minimization is defined in a similar way. Given the optimal value z of a pseudo-Boolean program max t subject to C, we can represent the solution set of the optimization problem by solving $C \cup \{t = z\}$ [Boc91].

3.1.2 Normal Forms

We classify pseudo-Boolean constraints and present for each kind of pseudo-Boolean constraints an appropriate normal form.

Non-linear Pseudo-Boolean Constraints

The most general class of pseudo-Boolean constraints are non-linear pseudo-Boolean constraints. In Section 3.1.1 we already have seen a normal form for pseudo-Boolean terms. We adopt this normal form to represent arbitrary non-linear pseudo-Boolean constraints. We distinguish between non-linear pseudo-Boolean equations and non-linear pseudo-Boolean inequalities.

It is sufficient to consider only non-linear pseudo-Boolean inequalities of the form $t_1 \geq t_2$, since the following equations obviously hold.

$$\begin{array}{rcl} \mathsf{Ext}(t_1 > t_2) & = & \mathsf{Ext}(t_1 \geq t_2 + 1) \\ \mathsf{Ext}(t_1 \leq t_2) & = & \mathsf{Ext}(t_2 \geq t_1) \\ \mathsf{Ext}(t_1 < t_2) & = & \mathsf{Ext}(t_2 \geq t_1 + 1) \end{array} \tag{3.3}$$

Next, we define a unique normal form for non-linear pseudo-Boolean inequalities.

Definition 3.1.8 A non-linear pseudo-Boolean inequality

$$\sum_{I \in \Omega} c_I \prod_{i \in I} X_i \geq d \ , \tag{3.4}$$

where Ω is a collection of subsets of $\{1, \ldots, n\} \neq \emptyset$ and coefficients $c_I \in \mathbb{Z} \setminus \{0\}$ is in *(non-linear pseudo-Boolean) normal form.*

Proposition 3.1.9 *For each non-linear pseudo-Boolean inequality there exists an equivalent non-linear pseudo-Boolean inequality in normal form.*

Proof. Let $t_1 \geq t_2$ be an arbitrary non-linear pseudo-Boolean inequality. Obviously, $t_1 \geq t_2$ is equivalent to $t_1 - t_2 \geq 0$. By Proposition 3.1.1 there is a unique normal form for the pseudo-Boolean term $t_1 - t_2$. For the left-hand side of the pseudo-Boolean inequality the only additional restriction is that we require the index sets $I \in \Omega$ to be non-empty. For an empty index set the pseudo-Boolean term $c_\emptyset \prod_{i \in \emptyset} X_i$ is simply $c_\emptyset$. By adding $-c_\emptyset$ on both sides of the non-linear pseudo-Boolean inequality we obtain the desired normal form. □

The non-linear pseudo-Boolean normal form of a pseudo-Boolean equation is defined in a similar way. Note that

$$\mathsf{Ext}(t_1 = t_2) = \mathsf{Ext}(\{t_1 \geq t_2, t_2 \geq t_1\})$$

and therefore non-linear pseudo-Boolean inequalities are sufficient to describe arbitrary non-linear constraints.

Linear Pseudo-Boolean Constraints

Linear pseudo-Boolean constraints are an important subclass of general pseudo-Boolean constraints. In most applications linear pseudo-Boolean constraints are sufficiently expressive to naturally formulate the problem. Linear 0-1 integer programming [NW88], a discipline of mathematics and operations research, deals with the problem of solving linear pseudo-Boolean programs. Furthermore, each non-linear pseudo-Boolean constraint is equivalent to a set of linear pseudo-Boolean constraints. We will present in Chapter 9 methods that transform a non-linear pseudo-Boolean constraint into an equivalent set of linear pseudo-Boolean constraints.

A linear pseudo-Boolean constraint is a pseudo-Boolean constraint between linear pseudo-Boolean terms. As for non-linear pseudo-Boolean constraints it is sufficient to consider only linear pseudo-Boolean inequalities of the form $t \geq d$, where t is a linear pseudo-Boolean term and d is an integer. We assume that all linear pseudo-Boolean terms are of the form

$$c + \sum_{i \in I} c_i \cdot L_i \ ,$$

where $c, c_i \in \mathbb{Z}$ and the L_i are literals for all $i \in I$. A product $c_i \cdot L_i$ is abbreviated by $c_i L_i$. Next, we define a normal form for linear pseudo-Boolean inequalities that allows us to detect immediately whether the pseudo-Boolean inequality is a tautology or a contradiction.

Definition 3.1.10 A linear pseudo-Boolean inequality $c_1 L_1 + \cdots + c_n L_n \geq d$ is in *(pseudo-Boolean) normal form* if

$$d \geq c_1 \geq \cdots \geq c_n \geq 1 \text{ and } \mathsf{Var}(L_i) \neq \mathsf{Var}(L_j) \text{ for all } 1 \leq i < j \leq n \ .$$

We assume that $d \geq 1$, since otherwise the linear pseudo-Boolean inequality in normal form is a tautology and therefore need not be considered.

Proposition 3.1.11 *[HR68] For each linear pseudo-Boolean inequality, which is not a tautology, there exists an equivalent linear pseudo-Boolean inequality in normal form.*

Proof. We constructively describe how to obtain the pseudo-Boolean normal form of a linear pseudo-Boolean inequality. We begin with an arbitrary linear pseudo-Boolean inequality

$$e_1L'_1 + \cdots + e_mL'_m \geq d' \ . \tag{3.5}$$

First, we apply several arithmetic equivalence transformations. We rewrite (3.5) such that literals containing the same variable are grouped together and obtain

$$a_1X_1 + b_1\overline{X_1} + \cdots + a_nX_n + b_n\overline{X_n} \geq d' \ ,$$

where the X_i are pairwise different. For each i such that $a_i = b_i$ we can simplify $a_iX_i + b_i\overline{X_i}$ to the constant a_i and move the constant a_i to the right-hand side. So let us assume that $a_i \neq b_i$ for all $1 \leq i \leq n$. Next, we replace $a_iX_i + b_i\overline{X_i}$ by $c'_iL_i + c''_i$ for all $1 \leq i \leq n$ according to

$$c'_iL_i + c''_i := \begin{cases} (a_i - b_i)X_i + b_i & \text{if} \quad a_i > b_i \\ (b_i - a_i)\overline{X_i} + a_i & \text{if} \quad b_i > a_i \ . \end{cases}$$

Note that the c'_i are all positive. Bringing the constants c''_i to the right-hand side gives us the new right-hand side $d = d' - \sum_{i=1}^{n} c''_i$. After re-indexing according to the ordering restriction we have brought the linear 0-1 inequality into the form

$$c'_1L_1 + \cdots + c'_nL_n \geq d \ , \tag{3.6}$$

where $c'_1 \geq \cdots \geq c'_n \geq 1$ and $\mathsf{Var}(L_i) \neq \mathsf{Var}(L_j)$ for all $1 \leq i < j \leq n$. Note that $d \geq 1$, since otherwise (3.5) is a tautology. So far we have only applied arithmetic equivalence transformations, hence an assignment α satisfies (3.5) if and only if α satisfies (3.6). When constructing the pseudo-Boolean normal form of a linear pseudo-Boolean inequality we can detect at this point whether we have a tautology or not.

Suppose that $c'_i > d$ for some i, then every assignment α with $\alpha(L_i) = 1$, maps the left-hand side of (3.6) to an integer greater than d and satisfies (3.6). For all assignments α with $\alpha(L_i) = 0$, the value of the left-hand side is independent of c'_i. Hence, we can safely replace each c'_i by d if $c'_i > d$. Formally, we define

$$c_i := \begin{cases} c'_i & \text{if} \quad c'_i \leq d \\ d & \text{if} \quad c'_i > d \end{cases} \tag{3.7}$$

for all $1 \leq i \leq n$, and thereby obtain the pseudo-Boolean normal form

$$c_1 L_1 + \cdots + c_n L_n \geq d \tag{3.8}$$

of (3.5) with $d \geq c_1 \geq \cdots \geq c_n \geq 1$. Obviously, an assignment α satisfies (3.5) if and only if α satisfies (3.6) if and only if α satisfies (3.8). □

The last step of the normalization process as described by (3.7) is also called *coefficient reduction* [CJP83].

We abbreviate a linear pseudo-Boolean inequality in normal form $c_1 L_1 + \cdots + c_n L_n \geq d$ by $cL \geq d$ and view cL as a set of products $\{c_i L_i \mid 1 \leq i \leq n\}$. For a set of products cL we denote by $\sum c := \sum_{i=1}^{n} c_i$ the sum over the coefficients of cL. Note that a linear pseudo-Boolean inequality in normal form $cL \geq d$ is satisfiable if and only if $\sum c \geq d$, because $c_i > 0$ for all $1 \leq i \leq n$. Hence, $cL \geq d$ is unsatisfiable if and only if $\sum c < d$, and we can easily decide whether $cL \geq d$ is $\bot$.

Example 3.1.12 Let us transform the linear pseudo-Boolean inequality

$$\begin{array}{l} -6 \cdot X_1 + -5 \cdot X_6 + 4 \cdot X_3 + 3 \cdot X_6 + 3 \cdot X_4 + \\ 3 \cdot X_3 + 2 \cdot X_1 + 2 \cdot \overline{X_3} + 2 \cdot X_5 + -2 \cdot \overline{X_6} + 1 \cdot X_2 \geq 7 \end{array} \tag{3.9}$$

into pseudo-Boolean normal form using Proposition 3.1.11. First, we regroup the products and obtain

$$\begin{array}{l} 4 \cdot X_3 + 3 \cdot X_3 + 2 \cdot \overline{X_3} + -6 \cdot X_1 + 2 \cdot X_1 + \\ 3 \cdot X_4 + 2 \cdot X_5 + 1 \cdot X_2 + -5 \cdot X_6 + 3 \cdot X_6 + -2 \cdot \overline{X_6} \geq 7 \ , \end{array}$$

which simplifies to

$$\begin{array}{l} 7 \cdot X_3 + 2 \cdot \overline{X_3} + -4 \cdot X_1 + 0 \cdot \overline{X_1} + 3 \cdot X_4 + 0 \cdot \overline{X_4} + \\ 2 \cdot X_5 + 0 \cdot \overline{X_5} + 1 \cdot X_2 + 0 \cdot \overline{X_2} + -2 \cdot X_6 + -2 \cdot \overline{X_6} \geq 7 \ . \end{array}$$

Because $a_6 = -b_6 = -2$, we simplify $-2 \cdot X_6 + -2 \cdot \overline{X_6}$ to -2, and derive

$$\begin{array}{l} 7 \cdot X_3 + 2 \cdot \overline{X_3} + -4 \cdot X_1 + 0 \cdot \overline{X_1} + 3 \cdot X_4 + 0 \cdot \overline{X_4} + \\ 2 \cdot X_5 + 0 \cdot \overline{X_5} + 1 \cdot X_2 + 0 \cdot \overline{X_2} \geq 7 + 2 = 9 \ . \end{array}$$

Next, we replace the $a_i X_i + b_i \overline{X_i}$ by $c_i' L_i + c_i''$ according to Proposition 3.1.11 and obtain

$$5 \cdot X_3 + 2 + 4 \cdot \overline{X_1} + -4 + 3 \cdot X_4 + 0 + 2 \cdot X_5 + 0 + 1 \cdot X_2 + 0 \geq 9 \ .$$

By bringing the constants to the right-hand side we have

$$5 \cdot X_3 + 4 \cdot \overline{X_1} + 3 \cdot X_4 + 2 \cdot X_5 + 1 \cdot X_2 \geq 9 - 2 + 4 = 11 \ .$$

Since all coefficients are less than or equal to the right-hand side, coefficient reduction does not apply. After re-indexing we obtain the pseudo-Boolean normal form

$$5 \cdot L_1 + 4 \cdot L_2 + 3 \cdot L_3 + 2 \cdot L_4 + 1 \cdot L_5 \geq 11 \qquad (3.10)$$

of (3.9) with $L_1 = X_3$, $L_2 = \overline{X_1}$, $L_3 = X_4$, $L_4 = X_5$, and $L_5 = X_2$.

Assumption 3.1.13 From now on we assume that all linear pseudo-Boolean inequalities are in pseudo-Boolean normal form. Furthermore, we assume that in all sets of linear pseudo-Boolean inequalities each of the linear pseudo-Boolean inequalities in the set is in pseudo-Boolean normal form.

Next, we show how to derive valid linear pseudo-Boolean inequalities from a given one. A linear pseudo-Boolean inequality $cL \geq d$ *reduces* to $c'L \geq d'$ (the *reduction*) if $0 \leq c'_i \leq c_i$ for all $1 \leq i \leq n$ and $d' = d - (\sum c - \sum c')$. We require that $d' \geq 1$ and hence forbid reduction to tautologies. Because $0 \leq \alpha(cL) - d \leq \alpha(c'L) - d'$ for all solutions α, a linear pseudo-Boolean inequality dominates all its reductions. We denote by $\mathsf{Red}(cL \geq d)$ the set of all reductions of a linear pseudo-Boolean inequality $cL \geq d$. We say a linear pseudo-Boolean inequality $cL \geq d$ *strictly reduces* to $c'L' \geq d'$ if $c'L' \geq d'$ is a reduction and either $c'_i = c_i$ or $c'_i = 0$ for all $1 \leq i \leq n$. We say we have *eliminated* the literal L_i if $c'_i = 0$. We denote by $\mathsf{SRed}(cL \geq d)$ the set of all strict reductions of a linear pseudo-Boolean inequality $cL \geq d$. Note that $cL \geq d \in \mathsf{SRed}(cL \geq d)$.

Extended Clauses and Classical Clauses

A linear pseudo-Boolean inequality $cL \geq d$ whose coefficients are all 1 is called an *extended clause* and abbreviated by $L \geq d$. A *classical clause* is an extended clause with right-hand side $d = 1$.

The problem of whether a set of classical clauses is satisfiable (SAT) has been extensively studied in the past [DP60, vG84, MS85, GU89, JW90, HHT94]. It was the first problem proven to be NP-complete [Coo71, Kar72, GJ79]. Extended clauses are a generalization of classical clauses, where one of the main properties, namely domination, remains easily decidable. Note that a reduction of an extended clause $L \geq d$ is $L \setminus L' \geq d - |L'|$ with $L' \subseteq L$.

Lemma 3.1.14 *$L \geq d$ dominates $L' \geq d'$ if and only if*

$$|L \setminus L'| \leq d - d' \; . \tag{3.11}$$

Proof.

"$\Leftarrow$": First, we compute a reduction of $L \geq d$ by eliminating all literals that are not in L', i.e., $L \setminus L'$. The reduction is $L \setminus (L \setminus L') \geq d - |L \setminus L'|$, which is equivalent to $L \cap L' \geq d - |L \setminus L'|$. Thus, $L \geq d$ dominates $L \cap L' \geq d - |L \setminus L'|$. Since $|L \setminus L'| \leq d - d'$, we have $0 \leq d' \leq d - |L \setminus L'| \leq d$ and therefore $L \geq d$ also dominates $L \cap L' \geq d'$. Since $L \cap L' \subseteq L'$, we know that $L \cap L' \geq d'$ dominates $L' \geq d'$. Hence, $L \geq d$ dominates $L' \geq d'$.

"$\Rightarrow$": We assume that $|L \setminus L'| > d - d'$ and show that in that case $L \geq d$ does not dominate $L' \geq d'$ by constructing an assignment α satisfying $L \geq d$, but not $L' \geq d'$. We choose $\alpha \in \mathrm{Ext}(L \geq d)$ such that $\alpha(L) = d$ and $\alpha(L_i) = 0$ for all $L_i \in L' \setminus L$. Then we know that $\alpha(L') = d - |L \setminus L'|$. Since $|L \setminus L'| > d - d'$, we have $d - |L \setminus L'| < d'$ and therefore $\alpha(L') < d'$. But then α is not in the extension of $L' \geq d'$. Thus, $L \geq d$ does not dominate $L' \geq d'$. □

When deciding domination between two classical clauses $L \geq 1$ and $L' \geq 1$, condition (3.11) reduces to $|L \setminus L'| \leq 0$, which is equivalent to $L \subseteq L'$; the usual condition for deciding domination (or implication) between classical clauses.

Note that deciding domination between two arbitrary linear pseudo-Boolean inequalities is an NP-complete problem[1]. So extended clauses are a restriction of linear pseudo-Boolean inequalities that make an important problem easily decidable. On the other hand, we can view extended clauses as a generalization of classical clauses, where we retain computational manageability while increasing expressiveness.

We say a set of extended clauses S is in *normal form* if no extended clause in S dominates another extended clause in S. Obviously, every set of extended clauses S can be easily brought into normal form by deleting all extended clauses that are dominated by other extended clauses in S.

[1] A linear pseudo-Boolean inequality $cL \geq d$ dominates another linear pseudo-Boolean inequality $c'L' \geq d'$ if and only if the minimum of $c'L'$ subject to $cL \geq d$ is greater than d'. So deciding domination between two linear pseudo-Boolean inequalities involves solving a knapsack problem [NW88].

No.	X_1	X_2	X_3	X_4	X_5	X_6	X_7
1	1	-	-	-	-	-	-
2	0	1	1	-	-	-	-
3	0	1	0	1	-	-	-
4	0	1	0	0	1	-	-
5	0	1	0	0	0	1	-
6	0	0	1	1	-	-	-
7	0	0	1	0	1	1	-
8	0	0	1	0	1	0	1
9	0	0	0	1	1	1	-

Table 3.1 Families of Solutions

3.2 PREVIOUS WORK

Solving pseudo-Boolean programs has been an active research area since the early works from Dantzig [Dan57], Fortet [For59, For60], and Hammer [HRR63, HR68]. We present in this section an overview of the work on solving pseudo-Boolean constraints having in mind their applicability in the context of constraint logic programming.

3.2.1 Boolean Methods

One of the early attempts to solve pseudo-Boolean problems was to explicitly construct the set of solutions and represent them in a table of *families of solutions* [Ham64, HR68]. For example, consider the linear pseudo-Boolean inequality in normal form [HR68]

$$9 \cdot X_1 + 7 \cdot X_2 + 5 \cdot X_3 + 4 \cdot X_4 + 3 \cdot X_5 + 2 \cdot X_6 + X_7 \geq 9 \ . \qquad (3.12)$$

The solutions of (3.12) can be represented as in Table 3.1. Each solution is implicitly given by taking one line of Table 3.1 and assigning each variable the value indicated in the table if it is 0 or 1 and arbitrary otherwise. For example, line 4 represents one family of solutions, i.e., the set of all assignments α such that $\alpha(X_1) = \alpha(X_3) = \alpha(X_4) = 0$, $\alpha(X_2) = \alpha(X_5) = 1$, and $\alpha(X_i)$ arbitrary for all $i \geq 6$.

For a linear pseudo-Boolean inequality in normal form constructing a table representing families of solutions is straightforward. We define a procedure `families` that constructs the families of solutions of a linear pseudo-Boolean inequality. We use the following non-standard conventions. By $cL = c_1L_1 + c'L'$ we express splitting the non-empty set of products cL such that c_1L_1 is a product with the largest coefficient in cL and $c'L'$ is $cL \setminus \{c_1L_1\}$. The first argument of `families` is a list, the empty list is denoted by [ ]. We denote by $A@B$ the list, where B is appended to A. We define

$$\texttt{families}(A, cL, d) := \begin{cases} [\texttt{fill(A)}] & \text{if} \quad d \leq 0 \\ [\,] & \text{if} \quad d > 0 \text{ and } cL = \emptyset \\ L_1@L_2 & \text{if} \quad d > 0, cL = c_1L_1 + c'L' \\ & \quad\quad L_1 := \texttt{families}(A@[1], c'L', d - c_i) \\ & \quad\quad L_2 := \texttt{families}(A@[0], c'L', d) \end{cases} ,$$

where `fill` is a function that fills a list with "-"-signs. Then $\texttt{families}([\,], cL, d)$ computes the table of families of solutions of the linear pseudo-Boolean inequality $cL \geq d$.

The main problem with representing the solutions of a linear pseudo-Boolean inequality as a table of families of solutions is that the table tends to blow up exponentially even on simple linear pseudo-Boolean inequalities, like extended clauses of the form

$$L_1 + \cdots + L_{2n} \geq n \ .$$

Constructing a table of families of solutions for a set of linear pseudo-Boolean inequalities is described in [HR68]. Maintaining a table of families of solutions for a set of pseudo-Boolean constraints is not advantageous in the context of constraint logic programming, since incrementality is hard to achieve and typically the size of the table forbids its use.

Hammer et al. [HR68, GH71] already proposed another way to represent the families of solutions. The key idea is to construct a Boolean formula that has the same set of satisfying solutions than the pseudo-Boolean constraint. Given the table of families of solutions, it is straightforward to give an equivalent Boolean formula. Since each family of solutions is characterized by one line where some variables are fixed, we immediately have a Boolean formula that has the same set of solutions as represented by that line. We take the conjunction of the variables that are fixed to 1 and their negation if they are fixed to 0. For the pseudo-Boolean inequality (3.12) we obtain

$$\begin{array}{l} X_1 \vee \\ (\overline{X_1} \wedge X_2 \wedge X_3) \vee \\ (\overline{X_1} \wedge X_2 \wedge \overline{X_3} \wedge X_4) \vee \\ (\overline{X_1} \wedge X_2 \wedge \overline{X_3} \wedge \overline{X_4} \wedge X_5) \vee \\ (\overline{X_1} \wedge X_2 \wedge \overline{X_3} \wedge \overline{X_4} \wedge \overline{X_5} \wedge X_6) \vee \\ (\overline{X_1} \wedge \overline{X_2} \wedge X_3 \wedge X_4) \vee \\ (\overline{X_1} \wedge \overline{X_2} \wedge X_3 \wedge \overline{X_4} \wedge X_5 \wedge X_6) \vee \\ (\overline{X_1} \wedge \overline{X_2} \wedge X_3 \wedge \overline{X_4} \wedge X_5 \wedge \overline{X_6} \wedge X_7) \vee \\ (\overline{X_1} \wedge \overline{X_2} \wedge \overline{X_3} \wedge X_4 \wedge X_5 \wedge X_6) \ . \end{array} \tag{3.13}$$

It was realized in [GH71] that the negated variables are not needed in order to preserve equivalence. A simplified equivalent formula to (3.13) is therefore

$$\begin{array}{l} X_1 \vee \\ (X_2 \wedge X_3) \vee \\ (X_2 \wedge X_4) \vee \\ (X_2 \wedge X_5) \vee \\ (X_2 \wedge X_6) \vee \\ (X_3 \wedge X_4) \vee \\ (X_3 \wedge X_5 \wedge X_6) \vee \\ (X_3 \wedge X_5 \wedge X_7) \vee \\ (X_4 \wedge X_5 \wedge X_6) \ . \end{array} \tag{3.14}$$

The Boolean formula 3.14 is called *resolvent* [GH71]. Similarly, we can compute a set of classical clauses that is equivalent to a linear pseudo-Boolean inequality $cL \geq d$. For this purpose we consider the negation of $cL \geq d$, namely $cL < d$, and bring it in pseudo-Boolean normal form. From the family of the solutions we construct again the resolvent. Since this resolvent is true if and only if the linear pseudo-Boolean inequality $cL \geq d$ is not satisfied, we obtain a Boolean formula that is equivalent to $cL \geq d$ by negating the obtained resolvent. With De Morgan's Law we obtain a conjunction of classical clauses. For the pseudo-Boolean inequality (3.12) we construct the following set of classical clauses.

$$\begin{array}{ll} \{ & X_1 + X_2 + X_3 + X_4 \geq 1, X_1 + X_2 + X_3 + X_5 \geq 1, \\ & X_1 + X_2 + X_3 + X_6 \geq 1, X_1 + X_2 + X_4 + X_5 \geq 1, \\ & X_1 + X_2 + X_4 + X_6 + X_7 \geq 1, X_1 + X_3 + X_4 + X_5 + X_6 \geq 1 \quad \} \end{array} \tag{3.15}$$

So apparently in [GH71] an algorithm was constructed generating a set of classical clauses[2] equivalent to a linear pseudo-Boolean inequality. The basic idea of *simplifying* the resulting set of clauses was also present in [GH71],

[2]The classical clauses generated by the resolvent method are so called *cover inequalities* and play today an important role in 0-1 integer programming [NW88].

although restricted to obvious simplifications like fixing variables and deleting dominated classical clauses. As a "promising idea" in the appendix of [GH71] "a B-B-B (Boolean branch-and-bound) type approach" is suggested for solving pseudo-Boolean optimization problems. Unfortunately, the method failed, mainly because of the large number of constraints in the equivalent set of classical clauses [GG80]. As already mentioned, reducing constraint solving of pseudo-Boolean constraints to constraint solving of classical clauses is interesting in the context of constraint logic programming, since

- SAT-procedures, deciding whether a set of clauses admits a satisfying assignment, have evolved recently [JW90, Zha93, HV94] and even efficient incremental SAT-procedures are available [Hoo93].
- there is an appropriate solved form, i.e., the set of prime implicates [JP90], and incremental algorithms for maintaining the solved form are available [KT90, Jac92].

We show that *extended clauses* provide the same good properties in the context of constraint logic programming as classical clauses. We develop methods transforming pseudo-Boolean constraints into an equivalent set of extended clauses and managing the solved form. The main advantage w.r.t. classical clauses is that typically a set of extended clauses is much more compact than an equivalent set of classical clauses.

3.2.2 Pseudo-Boolean Unification

The first constraint solver that embeds Boolean constraint-solving in a constraint logic programming language was based on *Boolean unification* [MN86, BS87, MN88, BR89, MN89]. Efficient implementations were possible using specialized data structures [Bry86, Kar88, Bry92].

Bockmayr [Boc91, Boc93b] generalizes Boolean unification to pseudo-Boolean unification. Since arbitrary pseudo-Boolean constraints can be formulated as a pseudo-Boolean unification problem, although more of theoretical interest, we consider as pseudo-Boolean constraints only pseudo-Boolean equations.

A ***pseudo-Boolean unification problem*** is a set of pseudo-Boolean equations

$$\{s_1 = t_1, \ldots, s_n = t_n\}$$

with $s_i, t_i \in \mathcal{T}(F_{\mathsf{psbool}}, \mathsf{Var}_{\mathsf{bool}})$ for all $1 \leq i \leq n$. A *pseudo-Boolean unifier* of a pseudo-Boolean unification problem is a substitution $\tau : \mathsf{Var}_{\mathsf{bool}} \to \mathcal{T}(F_{\mathsf{bool}}, \mathsf{Var}_{\mathsf{bool}})$ such that $\mathsf{Ext}(\tau(s_i)) = \mathsf{Ext}(\tau(t_i))$ for all $1 \leq i \leq n$. A *most general pseudo-Boolean unifier (mgu)* of a pseudo-Boolean unification problem is a substitution $\sigma : \mathsf{Var}_{\mathsf{bool}} \to \mathcal{T}(F_{\mathsf{bool}}, \mathsf{Var}_{\mathsf{bool}})$ such that for all pseudo-Boolean unifiers τ of the pseudo-Boolean unification problem there is a substitution $\lambda : \mathsf{Var}_{\mathsf{bool}} \to \mathcal{T}(F_{\mathsf{bool}}, \mathsf{Var}_{\mathsf{bool}})$ such that for all Boolean variables $X \in \mathsf{Var}_{\mathsf{bool}}$ the normal forms of the pseudo-Boolean terms $PB(\lambda(\sigma(X)))$ and $PB(\tau(X))$ are identical. Bockmayr [Boc93b] shows that pseudo-Boolean unification is unitary, which means that for any two unifiable pseudo-Boolean terms there is a most general pseudo-Boolean unifier, and gives a variable elimination procedure, which is a generalization of the Boolean unification algorithm of Büttner and Simonis [BS87] to the pseudo-Boolean case. We present another variable elimination procedure that needs not convert a pseudo-Boolean term into its Boolean equivalent.

It is sufficient to consider the pseudo-Boolean unification problem $t = 0$. First, suppose that t does not contain any Boolean variable, then the empty substitution is a mgu of t and 0 if the normal form t is identical to 0, otherwise t and 0 are not unifiable. Let t be a pseudo-Boolean term containing the Boolean variables $\{X_1, \ldots, X_{n+1}\}$. We can rewrite t into the form

$$t_1 \cdot X_{n+1} + t_2 \cdot \overline{X_{n+1}} \tag{3.16}$$

equivalent to t, where $t_1, t_2 \in \mathcal{T}(F_{\mathsf{psbool}}, \mathsf{Var}_{\mathsf{bool}})$ do not contain X_{n+1}. Constructing a mgu of t and 0 is possible, given a mgu of $t_1 = 0$ and $t_2 = 0$. First, we state an easy lemma.

Lemma 3.2.1 *[Boc91] Let t_1 and t_2 be pseudo-Boolean functions and s be a Boolean function. If $t_1 \cdot s + t_2 \cdot \overline{s} = 0$, then $t_1 = t_2 = 0$.*

Proof. Suppose that $t_1 \cdot s + t_2 \cdot \overline{s} = 0$. Multiplying this equation by s yields $t_1 \cdot s \cdot s + t_2 \cdot s \cdot \overline{s} = t_1 \cdot s = 0$. Similarly, multiplying with $\overline{s}$ yields $t_2 \cdot \overline{s} = 0$. Hence, $t_1 \cdot t_2 = t_1 \cdot t_2 \cdot (s + \overline{s}) = t_1 \cdot t_2 \cdot s + t_1 \cdot t_2 \cdot \overline{s} = 0 + 0 = 0$. Therefore, both $t_1 = 0$ and $t_2 = 0$. □

Next, we show how to a construct a mgu of (3.16) and 0, given a mgu of $t_1 = 0$ and $t_2 = 0$.

Proposition 3.2.2 *Let the pseudo-Boolean term t over the Boolean variables $\{X_1, \ldots, X_{n+1}\}$ be equivalent to $t_1 \cdot X_{n+1} + t_2 \cdot \overline{X_{n+1}}$ such that the Boolean* *variable X_{n+1} occurs in t but neither in t_1 nor in t_2. Let σ_1 be a mgu of $t_1 = 0$ and σ_2 be a mgu of $t_2 = 0$. Then the substitution σ defined by*

$$\begin{array}{rcll} \sigma(X_{n+1}) & := & Y & \text{where } Y \text{ is a new Boolean variable} \\ \sigma(X_i) & := & \sigma_1(X_i) \wedge Y \oplus \sigma_2(X_i) \wedge \overline{Y} & \text{for all } 1 \leq i \leq n \end{array} \tag{3.17}$$

is a most general pseudo-Boolean unifier of $t = 0$. If either t_1 or t_2 is not unifiable with 0, then the substitution σ defined by

$$\begin{array}{rcll} \sigma(X_{n+1}) & := & \begin{cases} 1 & \text{if } t_1 \text{ is unifiable with } 0 \\ 0 & \text{if } t_2 \text{ is unifiable with } 0 \end{cases} & \\ \sigma(X_i) & := & \begin{cases} \sigma_1(X_i) & \text{if } \sigma_1 \text{ is the mgu of } t_1 \text{ and } 0 \\ \sigma_2(X_i) & \text{if } \sigma_2 \text{ is the mgu of } t_2 \text{ and } 0 \end{cases} & \text{for all } 1 \leq i \leq n \end{array} \tag{3.18}$$

is a most general pseudo-Boolean unifier of $t = 0$. If neither t_1 nor t_2 is unifiable with 0, then t is not unifiable with 0.

Proof. Because of Lemma 3.2.1, we know that a mgu σ' of $t = 0$ has to satisfy $\sigma'(t_1 \cdot X_{n+1}) = \sigma'(t_2 \cdot \overline{X_{n+1}}) = 0$. If neither t_1 nor t_2 is unifiable with 0, then t is obviously not unifiable with 0. Let t_1, but not t_2, be unifiable with 0. Since t_2 is not unifiable with 0, we have $\sigma'(\overline{X_{n+1}}) = 0$. Therefore, $\sigma'(X_{n+1}) = 1$ and $\sigma'(t_1) = 0$. Hence, a mgu σ' of $t = 0$ must be a mgu of $t_1 = 0$, except that $\sigma'(X_{n+1}) = 1$. Thus, σ is a mgu of $t = 0$ by construction. Similarly, if t_2, but not t_1, is unifiable with 0.

In the remaining case t_1 and t_2 are unifiable with 0. First, we show that σ is a pseudo-Boolean unifier of $t = 0$. Since $\sigma(t) = \sigma(t_1) \cdot Y + \sigma(t_2) \cdot \overline{Y}$, we know by Lemma 3.2.1 that σ is pseudo-Boolean unifier of $t = 0$ if it is a pseudo-Boolean unifier of $t_1 \cdot Y = 0$ and $t_2 \cdot \overline{Y} = 0$. We show that σ is a pseudo-Boolean unifier of $t_1 \cdot Y = 0$. The proof that σ is a unifier of $t_2 \cdot \overline{Y} = 0$ is similar.

Let $\sum_{I \in \Omega} c_I \prod_{i \in I} X_i$ be the pseudo-Boolean normal form of t_1, then $t_1 \cdot Y$ is equivalent to

$$\sum_{I \in \Omega} c_I \prod_{i \in I} (X_i \cdot Y) \; .$$

Hence, $\sigma(t_1 \cdot Y)$ is equivalent to

$$\sum_{I \in \Omega} c_I \prod_{i \in I} ((\sigma_1(X_i) \wedge Y \oplus \sigma_2(X_i) \wedge \overline{Y}) \cdot Y)$$

by definition of σ. The inner term $(\sigma_1(X_i) \wedge Y \oplus \sigma_2(X_i) \wedge \overline{Y}) \cdot Y$ is equivalent to $\sigma_1(X_i) \cdot Y$. Thus, $\sigma(t_1 \cdot Y)$ is equivalent to

$$\sum_{I \in \Omega} c_I \prod_{i \in I} (\sigma_1(X_i) \cdot Y)$$

and therefore to $\sigma_1(t_1) \cdot Y$. But σ_1 is a pseudo-Boolean unifier of $t_1 = 0$ and thus of $t_1 \cdot Y = 0$.

It remains to show that σ is a mgu of $t = 0$. Let τ be an arbitrary pseudo-Boolean unifier of $t = 0$, then τ is a pseudo-Boolean unifier of $t_1 = 0$ and $t_2 = 0$. Since σ_1 is a mgu of $t_1 = 0$, we know that there is a substitution λ_1 such that $\lambda_1(\sigma_1(X_i)) = \tau(X_i)$ for all $1 \leq i \leq n$. Likewise, there is a substitution λ_2 such that $\lambda_2(\sigma_2(X_i)) = \tau(X_i)$ for all $1 \leq i \leq n$. We define the substitution λ by

$$\begin{array}{rcll} \lambda(Y') & := & \lambda_1(Y') \wedge \tau(X_{n+1}) \oplus \lambda_2(Y') \wedge \overline{\tau(X_{n+1})} & \text{for all } Y' \in V \\ \lambda(Y) & := & \tau(X_{n+1}) & \end{array}$$

where V is the set of all Boolean variables occurring in $\sigma_1(X_i)$ or $\sigma_2(X_i)$ for all $1 \leq i \leq n$. Similar to the proof of $\sigma(t_1) \cdot Y = \sigma_1(t_1) \cdot Y$, we can show that $\lambda(\sigma_1(X_i)) \wedge \lambda(Y) = \lambda_1(\sigma_1(X_i)) \wedge \lambda(Y)$ for all $1 \leq i \leq n$ and therefore $\lambda(\sigma_1(X_i)) \wedge \lambda(Y) = \tau(X_i) \wedge \lambda(Y)$ for all $1 \leq i \leq n$. Similarly, we have $\lambda(\sigma_2(X_i)) \wedge \overline{\lambda(Y)} = \tau(X_i) \wedge \overline{\lambda(Y)}$ for all $1 \leq i \leq n$. In summary, we have for all $1 \leq i \leq n$

$$\begin{array}{rcl} \lambda(\sigma(X_i)) & = & \lambda(\sigma_1(X_i)) \wedge \lambda(Y) \oplus \lambda(\sigma_2(X_i)) \wedge \overline{\lambda(Y)} \\ & = & \tau(X_i) \wedge \lambda(Y) \oplus \tau(X_i) \wedge \overline{\lambda(Y)} \\ & = & \tau(X_i) \end{array}$$

and $\lambda(\sigma(X_{n+1})) = \tau(X_{n+1})$ by definition. □

A variable elimination algorithm constructing a mgu of a pseudo-Boolean term t and 0 based on Proposition 3.2.2 is straightforward. We define

$$\begin{array}{lcl} \mathbf{vemgu}(0) & := & id \\ \mathbf{vemgu}(1) & := & \perp \\ \mathbf{vemgu}(t_1 \cdot X + t_2 \cdot \overline{X}) & := & \mathtt{combine}(\mathbf{vemgu}(t_1), \mathbf{vemgu}(t_2), X) \\[1ex] \mathtt{combine}(\perp, \perp, X) & := & \perp \\ \mathtt{combine}(\sigma, \perp, X) & := & \{X \mapsto 1, Z \mapsto \sigma(Z) \mid Z \neq X\} \\ \mathtt{combine}(\perp, \sigma, X) & := & \{X \mapsto 0, Z \mapsto \sigma(Z) \mid Z \neq X\} \\ \mathtt{combine}(\sigma_1, \sigma_2, X) & := & \{X \mapsto Y, Z \mapsto \sigma_1(Z) \cdot Y \oplus \sigma_2(Z) \cdot \overline{Y} \mid \\ & & \quad Z \neq X, Y \text{ and } Y \text{ is a new Boolean variable}\} \ , \end{array}$$

where *id* denotes the identity substitution. We assume that the argument term of **vemgu** is brought to the form $t_1 \cdot X \oplus t_2 \cdot \overline{X}$ if it contains a variable and to either 0 or 1 otherwise. Since the number of Boolean variables in the recursion of **vemgu** is strictly decreasing, **vemgu** is well defined. If t is unifiable with 0, then **vemgu**(t) is a mgu of $t = 0$, otherwise **vemgu**(t) is $\perp$. Correctness follows from Proposition 3.2.2.

In the context of constraint logic programming a mgu is an appropriate solved form. Global consistency is ensured and all possible solutions can be obtained by instantiating the new variables in the mgu. The algorithm can be made incremental as done for Boolean unification [BS87]. Problematic is the size of the mgu. Practical experience shows that even a small pseudo-Boolean unification problem can generate a very complex mgu. Another drawback is that deciding entailment of pseudo-Boolean constraints is not obvious, given a mgu.

3.2.3 Finite Domain Techniques

Local propagation methods for solving finite domain constraints [Mac77], introduced by van Hentenryck [vH89a] in the context of constraint logic programming and incorporated in the CHIP-system [DvHS+88a], have been successfully applied to many problems from operations research and artificial intelligence. Pseudo-Boolean constraints can be seen as finite domain constraints over the integers, where all constraint variables are restricted to the special domain $\{0,1\}$.

Solving Boolean constraints with finite domain techniques in CLP-systems is evaluated in [SD90, CD93]. The efficiency of the finite domain constraint solver on some typical Boolean benchmark problems is comparable to special purpose Boolean constraint solvers [CD93, CD94].

The idea of finite domain constraint solving is to reduce the domain of the variables whenever possible by the inference rules *forward checking* and *(partial) lookahead* [vH89a]. The only restriction possible when working with the restricted domain $\{0,1\}$ is to fix a Boolean variable either to 0 or to 1. Thus, domain reduction methods developed for arbitrary finite domains are too general for the domain $\{0,1\}$. The inference rules reduce to a consistency check described by the procedure `fixing` presented in Section 7.5 in the case of linear pseudo-Boolean inequalities. But, general finite domain techniques do not exploit the special structure of pseudo-Boolean constraints. In particular, non-linear finite domain constraints are typically suspended until they become linear and therefore are not actively used to prune the search space, whereas non-linear pseudo-Boolean constraints can be reformulated as an equivalent set of linear pseudo-Boolean inequalities with the method presented in Chapter 9. Moreover, powerful reformulation techniques, which are available for pseudo-Boolean constraints, are not yet fully investigated for finite domain constraint solvers based on local consistency techniques but can be used in order to enhance constraint solving [BB95].

The semantic problems of local consistency techniques in the context of constraint logic programming is discussed in Section 2.4 and we do not elaborate further on finite domain techniques for solving pseudo-Boolean constraints.

3.2.4 0-1 Integer Programming

Linear programming based methods are currently the most evolved tools for solving linear pseudo-Boolean optimization problems [NW88]. In particular, polyhedral cutting plane techniques have been successfully applied to a large variety of hard combinatorial problems [Gom58, Chv73, CJP83, BCC94]. The basic idea of solving pseudo-Boolean constraints with 0-1 integer programming methods is to solve first a *relaxation* of the original problem. Assuring local consistency in a finite domain solver can also be seen as solving a relaxation of the original problem. Unlike the obligatory enumeration in finite domain constraint solving, polyhedral cutting plane techniques can be used to generate valid inequalities that yield a stronger relaxation of the problem. A pure cutting plane algorithm reformulates the problem until the relaxation of the reformulated problem is sufficiently strong, i.e., until solving the relaxation solves the initial problem. In the context of constraint logic programming this means reformulating a set of constraints until satisfiability (and entailment) becomes easily decidable. Moreover, the reformulated problem serves as solved form. The use of cutting plane techniques in the context of constraint logic programming was first proposed by Bockmayr [Boc92a, Boc93a, Boc93c] and is currently an active research area [Boc94, Boc95]. After introducing the polyhedral view of a set of linear pseudo-Boolean inequalities and the relaxation of the problem, we review some basic results from polyhedral theory [NW88] and investigate the approach in the context of constraint logic programming [Boc94, Boc95].

In 0-1 integer programming a set of linear pseudo-Boolean inequalities is formulated as matrix

$$\begin{array}{cccc} a_{11}X_1 & +\cdots+ & a_{1n}X_n & \leq & b_1 \\ \vdots & & \vdots & & \vdots \\ a_{m1}X_1 & +\cdots+ & a_{mn}X_n & \leq & b_m \end{array} , \tag{3.19}$$

or shortly as $aX \leq b$ with $a \in \mathbb{Z}^{m \times n}$, $X \in \{0,1\}^n$ and $b \in \mathbb{Z}^m$. From a mathematical point of view $aX \leq b$ describes a possibly empty set of 0-1 points in $\mathbb{R}^n$. By replacing the condition $X \in \{0,1\}^n$ by $X \in [0,1]^n$, i.e.,

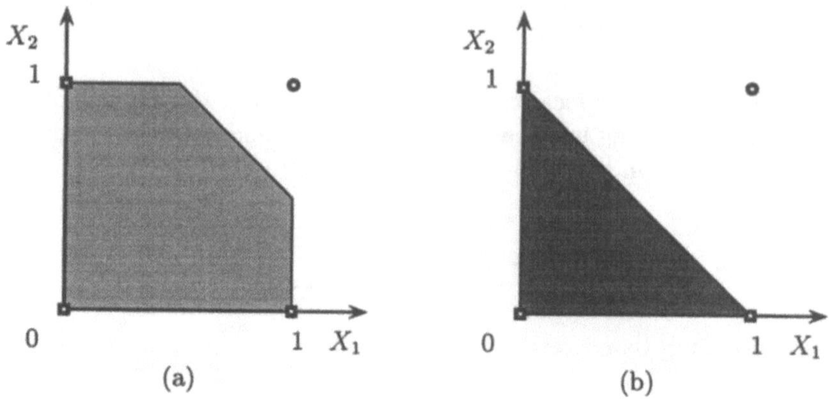

Figure 3.1 (a) Polyhedron of $2 \cdot X_1 + 2 \cdot X_2 \leq 3$ and $X_1, X_2 \in [0,1]$. (b) The convex hull of its 0-1 solutions.

$0 \leq X_i \leq 1$ for all $1 \leq i \leq n$, we obtain the *linear programming relaxation* of $aX \leq b$. Thus, system (3.19) becomes a system of linear arithmetic constraints over the real numbers. The set of points $X \in [0,1]^n$ satisfying (3.19) defines a *polytope* P in $\mathbb{R}^n$. The set of 0-1 points in the polytope is the extension of the original problem, i.e., $\text{Ext}(aX \leq b) = P \cap \{0,1\}^n$.

Example 3.2.3 Let $aX \leq b$ be

$$2 \cdot X_1 + 2 \cdot X_2 \leq 3 \ .$$

The corresponding polytope P is given in Figure 3.1(a) as gray region. The 0-1 solutions points $(0,0), (0,1), (1,0)$ lying in P are marked by ■ .

For a full-dimensional polytope $P \subseteq [0,1]^n$ there is up to multiplication of inequalities by positive real numbers a unique representation consisting of the set of all *facet defining inequalities* of P. A facet defining inequality of a polytope P is an inequality for which n affinely independent points in P are satisfied at equality. Note that there are several different polyhedra containing the same set of 0-1-solution points. A particularly strong polytope is the convex hull $conv(P \cap \{0,1\})$.

Example 3.2.4 Let $aX \leq b$ be $2 \cdot X_1 + 2 \cdot X_2 \leq 3$, as in Example 3.2.3, and P be the corresponding polytope. The convex hull of $P \cap \{0,1\}^n$ is the gray region in Figure 3.1(b). For example, an inequality description of this convex hull is $X_1 + X_2 \leq 1$ plus the trivial inequalities $0 \leq X_1, X_2 \leq 1$. Note that $conv(P \cap \{0,1\}^n) \subset P$.

Bockmayr [Boc94, Boc95] compares several kinds of solved forms for pseudo-Boolean constraints and shows that an inequality description of the convex hull of the 0-1 solutions points can be seen as (ideal) solved form in the context of constraint logic programming. The convex hull is empty if and only if the set of pseudo-Boolean inequalities is unsatisfiable and thus has a trivial inequality description. Furthermore, deciding entailment, given an inequality description of the convex hull, reduces to solving a linear program. Note that a linear pseudo-Boolean inequality $cX \leq d$ is entailed by a set of linear pseudo-Boolean inequalities $aX \leq b$ if and only if the maximum of cX subject to $aX \leq b$ is smaller than or equal to d. When solving the linear program $\max\{aX \mid a'X \leq b'\}$, where $a'X \leq b'$ is the inequality description of the non-empty convex hull of $aX \leq b, X \in \{0,1\}^n$, the optimal solution is ensured to be a 0-1 solution.

Enumeration and cutting plane generation are the standard techniques of linear programming based 0-1 integer programming for solving linear pseudo-Boolean optimization problems. A pure enumeration algorithm, *branch-and-bound*, solves first the linear programming relaxation of the problem. If the solution happens to be 0-1, then the problem is solved. Otherwise, a branching variable having a fractional value in solution of the linear programming relaxation, is selected and the two subproblems, where the branching variable is set to 0 and 1 respectively, are solved. The optimal solution is the better of the optimal solutions of the two subproblems. When a subproblem is solved, then all other subproblems whose solution of the linear programming relaxation is not as good as the current integer solution can be deleted, since they cannot contain the optimum. This step is also called bounding.

Instead of branching whenever a fractional solution of the current problem shows up, we can also generate a valid pseudo-Boolean inequality w.r.t. the 0-1 solutions that invalidates the current fractional solution. These valid inequalities are called *cutting planes*, or *cuts*. We say that an inequality $cX \leq d$ is *stronger* than $c'X \leq d'$ w.r.t. a set of points M, if each $X \in M$ satisfying $cX \leq d$ satisfies $c'X \leq d'$ [Boc94]. Generating strong cuts, ideally facet defining inequalities of the convex hull of the 0-1 solution points, may reduce the number of enumeration steps by an order of magnitude [PR91, CCT87]. Whether a cut is generated or branching takes place is typically decided by heuristics. Algorithms that combine cutting plane generation and enumeration are called *branch-and-cut*.

A pure cutting plane algorithm allows the generation of a cutting plane only if a fractional solution occurs. This process is repeated[3] until a 0-1 solution is reached or the problem is found to be infeasible. In other words, the original inequality description is reformulated and getting closer and closer to an inequality description of the convex hull of the 0-1 solutions of the problem. A cutting plane algorithm fits well the requirements imposed by constraint logic programming [Boc94].

- The algorithm is incremental in nature. Incremental algorithms for maintaining a set of linear inequalities in the context of constraint logic programming are available [JMSY90].
- The solved form provides an easy check of satisfiability and entailment.

The ideal solved form, i.e., the inequality description of the convex hull of the 0-1 solution points, may require an exponential number of inequalities. However, we can stop the algorithm if the problems imposed by the current state of the CLP-program are solved. The idea is to compute an approximation of the convex hull that allows one to decide satisfiability of the current set of constraints and answer the current queries about entailment. The approximation is computed by adding cutting planes and simplifying the inequality description. The advantage of this approach is that cutting planes are added only when needed and there is a constant progress towards the solved form. Bockmayr [Boc94] presents such a constraint solver for CLP($\mathcal{PB}$) based on the new and powerful cutting plane method *lift and project* [BCC93b, BCC93a].

In the next section we investigate a logical view of linear pseudo-Boolean inequalities and adopt the idea of generating valid inequalities in order to approximate a solved form.

3.3 GENERALIZED RESOLUTION

Generating strong valid inequalities w.r.t. $\mathbb{R}^n$ is the aim of the cutting plane methods presented in Section 3.2.4. The inequalities representing the facets of the convex hull of the 0-1 solutions points are strongest valid inequalities w.r.t. $\mathbb{R}^n$. Hooker [Hoo92a] investigates strong valid inequalities w.r.t. $\{0,1\}^n$ and presents a deductive system to generate them.

[3]Termination of this process can be enforced by several restrictions depending on the kind of cutting planes that are added. We refer to [NW88] for further explanation.

Definition 3.3.1 A linear pseudo-Boolean inequality $cL \geq d$ is *stronger* than $c'L' \geq d'$ w.r.t. a set of 0-1 points Sol, if

$$Sol \subseteq \mathsf{Ext}(cL \geq d) \subseteq \mathsf{Ext}(c'L' \geq d') \ .$$

A *logic cut* w.r.t. a set of linear pseudo-Boolean inequalities is a linear pseudo-Boolean inequality dominated by S, but not dominated by any linear pseudo-Boolean inequality in S.

Basic principles of linear programming based methods for 0-1 integer programming carry over to the "logical view" [Hoo94b] of linear pseudo-Boolean optimization problems. The linear programming relaxation is replaced by a discrete relaxation and cutting planes are replaced by logical implications. The nice properties of cutting plane algorithms still hold for logic cut algorithms. The two views of pseudo-Boolean problems do not exclude each other. Exploiting the structure of particular problems by generating cuts, either logic or polyhedral ones, is possible. Moreover, the logical view often provides a better intuitive way of finding strong logic cuts when the analysis of the polyhedron is far too complex [Hoo94b, Hoo94a]. Although strong cuts w.r.t. $[0.1]^n$ and $\{0,1\}^n$ are quite different concepts, there are parallels (see for example Section 8.2).

3.3.1 Prime Inequalities

We investigate strong inequalities w.r.t. $\{0,1\}^n$. The strongest inequalities w.r.t. $\{0,1\}^n$ are so-called *prime inequalities* [Hoo92a].

Definition 3.3.2 A linear pseudo-Boolean inequality $cL \geq d$ is *prime* w.r.t. a set of linear pseudo-Boolean inequalities S if S dominates $cL \geq d$, but S dominates no other linear pseudo-Boolean inequality that strictly dominates $cL \geq d$.

A prime inequality is a strongest possible linear pseudo-Boolean inequality in the sense that its extension properly contains that of no other valid linear pseudo-Boolean inequality. Unfortunately, a prime inequality is not necessarily unique, i.e., a prime inequality may be equivalent to a number of other prime inequalities. The notion of prime inequality can also be defined w.r.t. a class T of linear pseudo-Boolean inequalities. Extended clauses are a class of linear pseudo-Boolean inequalities for which a unique set of prime inequalities exists.

Definition 3.3.3 An extended clause $L \geq d$ is *prime* w.r.t. a set of linear pseudo-Boolean inequalities S if S dominates $L \geq d$, but S dominates no other extended clause strictly dominating $L \geq d$.

A prime extended clause is necessarily unique, since different extended clauses necessarily have different extensions. We propose to regard *the equivalent set of all prime extended clauses as solved form of a set of pseudo-Boolean constraints.* This solved form satisfies the requirements imposed by constraint solving in the context of constraint logic programming.

Proposition 3.3.4 *Let S be a set of linear pseudo-Boolean inequalities. An extended clause $L \geq d$ is dominated by S if and only if $L \geq d$ is dominated by a prime extended clause w.r.t. S.*

Proof.

$\Leftarrow$: If $L \geq d$ is dominated by a prime extended clause w.r.t. S, then obviously $L \geq d$ is dominated by S.

$\Rightarrow$: Let S dominate $L \geq d$. It remains to show that there is a prime extended clause w.r.t. S that dominates $L \geq d$. If $L \geq d$ is prime w.r.t. S, then the proposition holds, since each extended clause dominates itself. Let $L \geq d$ be not prime w.r.t. S. Then, by Definition 3.3.3, there exists an extended clause $L' \geq d'$ dominated by S that strictly dominates $L \geq d$. Suppose that there is no prime extended clause w.r.t. S dominating $L \geq d$. Then there is no prime extended clause w.r.t. S dominating $L' \geq d'$, since it would also dominate $L \geq d$. Therefore, we have an extended clause strictly dominating $L \geq d$ for which there is no prime extended clause w.r.t. S strictly dominating it. We conclude that there must be an infinite chain of different extended clauses, each of them strictly dominating its predecessor. Therefore, there is an infinite chain of extensions such that each extension is a proper subset of its predecessor. Since any extension is finite, there cannot be such a chain. Hence, $L \geq d$ is dominated by a prime extended clause w.r.t. S. □

By Proposition 3.3.4 the basic functionalities of a pseudo-Boolean constraint solver, i.e., entailment and consistency check, are supported, when choosing the set of all prime extended clauses as solved form.

- An extended clause is dominated by a set of pseudo-Boolean constraints if and only if it is dominated by one prime extended clause of the set of pseudo-Boolean constraints.
- A set of pseudo-Boolean constraints is unsatisfiable if and only if its solved form is $\{\perp\}$. Note that $\perp$ dominates all extended clauses and that $\perp$ is dominated if and only if the set of pseudo-Boolean constraints is unsatisfiable.

Since deciding domination between extended clauses is easy (see Lemma 3.1.14), we have an easy check of entailment. In order to decide whether an arbitrary pseudo-Boolean constraint is dominated by a set of pseudo-Boolean constraints, we have to reformulate these constraints into an equivalent set of extended clauses. Techniques for such a reformulation are presented in Chapter 7 and Chapter 9.

The notion of prime inequalities and facet defining inequalities of the convex hull of a 0-1 solution set are not equivalent. A prime inequality need not be facet defining and a facet defining inequality may fail to be prime.

Example 3.3.5 [Hoo92a] Let S be the set of linear pseudo-Boolean inequalities

$$\{X_1 + X_2 \geq 1, X_1 + X_3 \geq 1\} \ .$$

The extension of S is given in the following table.

	α_1	α_2	α_3	α_4	α_5
X_1	1	1	0	1	1
X_2	0	0	1	1	1
X_3	0	1	1	0	1

The linear pseudo-Boolean inequality $X_1 + X_2 \geq 1$ is facet defining but not prime, since it is strictly dominated by the valid linear pseudo-Boolean inequality $2 \cdot X_1 + X_2 + X_3 \geq 2$. Note that the point (0,1,0) satisfies the former inequality but not the latter. On the other hand, the latter inequality is not facet defining although it is prime.

3.3.2 The Deductive System

Having defined the solved form, we are now interested in how to obtain it. Hooker [Hoo92a] presents the deductive system *Generalized Resolution* that generates a set of prime inequalities given a set of linear pseudo-Boolean inequalities, including a specialized version working on sets of extended clauses.

Definition 3.3.6 The deductive system *Generalized Resolution* consists of the transition rules, *resolution, diagonal sum,* and *simplify.* We denote by $\overline{L} := \{\overline{L_i} \mid L_i \in L\}$ the set of all negated literals of a set of literals L. Let S be a set of linear pseudo-Boolean inequalities.

Resolution : Add an extended clause $L \geq 1$, a *resolvent* of S, to S if

- [a] Each of the classical clauses $L' \geq 1$ and $L'' \geq 1$ is dominated by a linear pseudo-Boolean inequality in S,
- [b] $L' \cap \overline{L''} = \{L_i\}$, and
- [c] $L = (L' \cup L'') \setminus \{L_i, \overline{L_i}\}$.

Diagonal Sum Rule : Add a linear pseudo-Boolean inequality of the form $\sum_{1 \leq i \leq n} c_i L_i \geq d + 1$, a *diagonal sum* of S, to S if

- [a] for all $1 \leq j \leq n$ the linear pseudo-Boolean inequality $(c_j - 1)L_j + \sum_{1 \leq i \neq j \leq n} c_i L_i \geq d$, which is a reduction of $\sum_{1 \leq i \leq n} c_i L_i \geq d + 1$, is dominated by a linear pseudo-Boolean inequality in S,
- [b] there is no linear pseudo-Boolean inequality in S that dominates $\sum_{1 \leq i \leq n} c_i L_i \geq d + 1$, and
- [c] $\mathsf{Var}(L_i) \neq \mathsf{Var}(L_j)$ for all $i \neq j$.

Simplify : Delete a linear pseudo-Boolean inequality in S if it is dominated by another linear pseudo-Boolean inequality in S.

A resolvent is a resolvent in the classical sense (see Section 7.8 and 8.3 for a more general notion of resolvents), but we forbid construction of tautologies. The diagonal sum rule has no obvious relation to classical resolution. Intuitively, its role is to increase the coefficients and the right-hand side. We illustrate the diagonal sum rule by an example.

Example 3.3.7 [Hoo92a] Let S contain the following 4 linear pseudo-Boolean inequalities.

$$\begin{array}{rcrcrcrcr}
L_1 & + & 5 \cdot L_2 & + & 3 \cdot L_3 & + & L_4 & \geq & 4 \\
2 \cdot L_1 & + & 4 \cdot L_2 & + & 3 \cdot L_3 & + & L_4 & \geq & 4 \\
2 \cdot L_1 & + & 5 \cdot L_2 & + & 2 \cdot L_3 & + & L_4 & \geq & 4 \\
2 \cdot L_1 & + & 5 \cdot L_2 & + & 3 \cdot L_3 & & & \geq & 4
\end{array}$$

Note that all linear pseudo-Boolean inequalities are identical, except that the coefficients of the diagonal term, highlighted by the box, are reduced by one. The diagonal sum rule states that the linear pseudo-Boolean inequality obtained by taking a coefficient/literal pair with the maximal coefficient of each column for the left-hand side and taking the right-hand side plus one, is a valid linear pseudo-Boolean inequality. In our example we derive that

$$2 \cdot L_1 + 5 \cdot L_2 + 3 \cdot L_3 + L_4 \geq 5$$

is dominated by S.

Hooker [Hoo92a] shows that the transition rules of *Generalized Resolution* generate a complete set of prime inequalities.

Theorem 3.3.8 *[Hoo92a] The repeated application of the transition rules* **Resolution, Diagonal Sum,** *and* **Simplify** *on a set of linear pseudo-Boolean inequalities S until none of the rules is applicable yields a complete set of prime inequalities of S.*

Proof. Can be found in [Hoo92a](pp. 277–278). □

It is interesting to note that the two transition rules resolution and diagonal sum generate a subclass of so-called *Chvátal-Gomory cutting planes* [Gom58], which are used in linear programming based 0-1 integer programming. A Chvátal-Gomory cutting plane is obtained from a non-negative linear combination of a set of linear pseudo-Boolean inequalities followed by integer rounding [NW88]. Let

$$r_1 L_1 + \cdots + r_n L_n \geq s \tag{3.20}$$

be a linear pseudo-Boolean inequality obtained by a non-negative linear combination of a set S of linear pseudo-Boolean inequalities, where the r_i and s may be fractional but greater zero. The linear pseudo-Boolean inequality (3.20) is obviously valid. We round up the fractional coefficients of the left-hand side yielding

$$\lceil r_1 \rceil L_1 + \cdots + \lceil r_n \rceil L_n \geq s$$

with $\lceil a \rceil = \min(\{b \in \mathbb{N} \mid b \geq a\})$. Because rounding only increases the left-hand side, we have generated a valid inequality w.r.t. S. Since the coefficients $\lceil r_i \rceil$ are integer and $L_i \in \{0,1\}$ for all $1 \leq i \leq n$, the left-hand side is integer for all assignments. Therefore, the right-hand side s can be round up to the next integer, i.e to $\lceil s \rceil$. We obtain the valid linear pseudo-Boolean inequality

$$\lceil r_1 \rceil L_1 + \cdots + \lceil r_n \rceil L_n \geq \lceil s \rceil \ ,$$

dominated by S. Next, we show that the resolution and diagonal sum rule generate Chvátal-Gomory cutting planes.

Let $(L \cup L') \setminus \{L_i, \overline{L_i}\} \geq 1$ be the resolvent of the two classical clauses $L \geq 1$ and $L' \geq 1$. The sum, a linear combination, of the classical clauses is $L + L' \geq 2$. The subterm $L_i + \overline{L_i}$ can be simplified to 1, which is brought to the right-hand side. Because only one literal occurs positively in one and negatively in the other classical clause, we know that $L + L' \geq 2$ is equivalent to

$$2(L \cap L') + 1((L \cup L') \setminus ((L \cap L') \cup \{L_i, \overline{L_i}\}) \geq 1 \ .$$

We divide this inequality[4] by 2 and after integer rounding we obtain

$$1(L \cap L') + 1((L \cup L') \setminus ((L \cap L') \cup \{L_i, \overline{L_i}\}) \geq 1 \ ,$$

which simplifies to the resolvent $(L \cup L') \setminus \{L_i, \overline{L_i}\} \geq 1$.

For the diagonal sum rule we consider the weighted reductions

$$(c_j - 1)L_j + \sum_{1 \leq i \neq j \leq n} c_i L_i \geq d - 1 \qquad | \cdot c_j/(W-1) \qquad \text{for all } 1 \leq j \leq n \quad (3.21)$$

of the diagonal sum $\sum_{1 \leq i \leq n} c_i L_i \geq d + 1$ with $W := \sum_{i=1}^{n} c_i$. The sum over all the weighted reductions (3.21) is then

$$\sum_{i=1}^{n} c_i L_i \geq (d-1)W/(W-1) \ . \qquad (3.22)$$

Since $W \geq d$, otherwise the linear pseudo-Boolean inequality is unsatisfiable, we have

$$d - 1 < (d-1)W/(W-1) \leq d \ .$$

Thus, $\lceil (d-1)W/(W-1) \rceil = d$ and integer rounding of (3.22) yields the desired diagonal sum.

As for polyhedral cutting planes it is not necessary that the logic cuts are generated by a specific method. Any method generating valid linear pseudo-Boolean inequalities can be used, if we assure that all possible resolvents and diagonal sums, or linear pseudo-Boolean inequalities dominating them, are generated. For extended clauses we discuss in detail how to generate "good logic cuts" in Chapter 6 and Chapter 8. The performance of our constraint solver depends heavily on the logic cut generation process. Studying the generation of logic cuts is a new promising field that combines operations research and theorem proving.

[4] Since the right-hand side is 1, coefficient reduction can be applied. Thus, bringing the sum into normal form also yields the resolvent.

3.3.3 Logic Cuts in Constraint Logic Programming

In the context of constraint logic programming a pure logic cut algorithm has the same advantages as a pure polyhedral cutting plane algorithm. By choosing extended clauses as a class of linear pseudo-Boolean inequalities for the solved form, we have a user-friendly and natural description of the 0-1 solutions. In order to cope with the exponential blow up in the number of extended clauses that are needed to represent the 0-1 solutions, we show how to relax our solved form such that just satisfiability and entailment can be decided easily. A pure logic cut algorithm is suitable in the context of constraint logic programming, since:

- The algorithm is incremental in its nature. Adding more information and asking more questions about entailment gives a constant progress versus the solved form, the set of prime inequalities.
- When choosing extended clauses as basic constraints, there is a unique normal form that allows for an easy test of satisfiability and entailment.

We believe that logic cut algorithms can be as efficient as polyhedral cutting plane methods and that many constraint sets that arise while executing constraint logic programs are better thought of logically.

Logic-based methods for solving pseudo-Boolean constraints in the context of constraint logic programming constitute a constraint solver for CLP($\mathcal{PB}$) providing global consistency and a user-friendly solved form. With the logic view of pseudo-Boolean constraints, we merge polyhedral techniques with automated theorem proving. Global consistency and support for entailment make 'real' CLP-programs handling combinatorial problems possible. To this end we present in the following chapters a collection of methods, including a pure logic cut algorithm and several reformulation and simplification techniques, whose combination result in a constraint solver for pseudo-Boolean constraints suitable in the context of constraint logic programming. We contribute to the study of logic cuts and present several logic cut generation and reformulation techniques with an emphasis on their applicability in the context of constraint logic programming.

4

A LOGIC CUT BASED CONSTRAINT SOLVER

We outline a constraint solver for pseudo-Boolean constraints based on the generation of logic cuts. The principle idea is to *add logic cuts to the current set of constraints as long as an appropriate relaxation of the problem is not able to answer the questions imposed by the current state of the* CLP*-program.*

4.1 THE SOLVED FORM

Constraint solving in the context of constraint logic programming means reformulating a set of pseudo-Boolean constraints into a solved form such that the solution space remains invariant. Unsatisfiability of a set of pseudo-Boolean constraints must be detected and the solved form should ease deciding entailment. For a set of pseudo-Boolean constraints we propose to regard its set of all prime extended clauses, as in Definition 3.3.3, as a solved form [Bar93b]. Providing the fully solved form at each step of computation is useless, since it may change completely while the constraint logic program proceeds. It is sufficient if the constraint solver is able to answer the satisfiability and entailment questions imposed by the current state of the CLP-program.

Since we assume that non-linear pseudo-Boolean constraints have been linearized in advance, using the techniques given in Chapter 9, we consider only linear pseudo-Boolean inequalities as pseudo-Boolean constraints. The current set of constraints, the *constraint store* [Sar93], then consists of a set of linear pseudo-Boolean inequalities. Because our ideal solved form consists of a set of extended clauses, it is convenient to separate the extended clauses from the set of linear pseudo-Boolean inequalities. Hence, the constraint store consists of a set of extended clauses E and a set of linear pseudo-Boolean inequalities I, written (I, E), and is equivalent to $I \cup E$.

Ideally, the set of linear pseudo-Boolean inequalities is empty and the set of extended clauses is the set of all prime extended clauses of the current constraint set. Since the number of all prime extended clauses of a set of pseudo-Boolean constraints may be exponentially large, we relax the solved form, at least while the computation continues. We do **not** relax the satisfiability test, and require that if the current set of pseudo-Boolean constraints is unsatisfiable, then the constraint store is $(\emptyset, \{\bot\})$.

If the current satisfiable constraint store is (I, E), we allow in principle any (I', E') such that $\mathsf{Ext}(I \cup E) = \mathsf{Ext}(I' \cup E')$. We only require that progress is made versus the solved form.

4.2 REACHING THE SOLVED FORM

Adding a new constraint to the constraint store gives rise to further reformulation of the constraint store. Satisfiability of the enlarged constraint set needs to be ensured and the entailment questions have to be answered[1].

First, we reformulate each pseudo-Boolean constraint that is added to the constraint store into an equivalent set of linear pseudo-Boolean inequalities and then add this set of linear pseudo-Boolean inequalities to the constraint store. Then we have to decide satisfiability of the current constraint store. A variant of the Davis-Putnam implicit enumeration method generalized for handling linear pseudo-Boolean inequalities, which is explained in detail in Chapter 5, serves for this purpose. The Davis-Putnam method for linear pseudo-Boolean inequalities is tailored to generate strong logic cuts in Chapter 6 as long as satisfiability is not ensured. For reasons of efficiency, we develop sophisticated simplification techniques that reformulate a current constraint store (I, E) into (I', E') such that $\mathsf{Ext}(I \cup E) = \mathsf{Ext}(I' \cup E')$, but (I', E') is in some sense *simpler* or *closer to the solved form*. Among these techniques we mainly distinguish between techniques working on linear pseudo-Boolean inequalities and techniques working only on extended clauses. The principle idea is to generate strong logic cuts that are not yet dominated by any single pseudo-Boolean constraint in the constraint store.

[1] In a traditional CLP-system, employing a standard depth-first search, entailment questions do not arise.

Since our ultimate goal is a solved form consisting only of extended clauses, we need to transform a linear pseudo-Boolean inequality into an equivalent set of extended clauses. We develop an efficient transformation method in Chapter 7. Although an exponential number of extended clauses may be generated, typically fewer extended clauses than classical clauses are needed to represent a linear pseudo-Boolean inequality. Moreover, we investigate the concept of symmetries in linear pseudo-Boolean inequalities that speeds up the transformation method and suggests a more compact representation of a set of linear pseudo-Boolean inequalities. Additionally, the method is refined in order to suit the needs of constraint logic programming and to avoid the exponential blow up in the number of extended clauses that may be necessary for representing a linear pseudo-Boolean inequality. We present techniques, like coefficient reduction and the generalization of the transformation method to constrained transformation, that *strengthen* linear pseudo-Boolean inequalities w.r.t. the current constraint store. Such a strengthened linear pseudo-Boolean inequality can also be seen as a logic cut, since the strengthened linear pseudo-Boolean inequality typically has fewer 0-1 solutions than the original one. We also show how to split linear pseudo-Boolean inequalities into several smaller ones when a complete transformation is unsuitable because of excessive storage requirements. The smaller linear pseudo-Boolean inequalities are more easily reformulated and strengthened. Furthermore, partially transforming a linear pseudo-Boolean inequality into an equivalent set of extended clauses is needed when deciding entailment of a linear pseudo-Boolean inequality.

The transformation techniques of Chapter 7 push information into the set of extended clauses. The efficiency of the constraint solver depends highly on a powerful simplification of the set of extended clauses. We present efficient simplification methods for extended clauses in Chapter 8. We restrict the application of the inference rules of *Generalized Resolution* [Hoo92a] to the production of stronger extended clauses, i.e., strong logic cuts, and show in detail how to obtain them efficiently.

Whenever a Boolean variable does not appear in the top level goal and is not visible in subsequent goals, we can project it away by the techniques presented in Chapter 10. At the latest, when the computation ends and the answer constraint has to be prepared, projecting away Boolean variables that do not occur in the top level goal is required.

In practice it is useful to maintain the fixed literals separately and ensure that none of the fixed literals occur in the remaining parts of the constraint store. Furthermore, equations between Boolean variables have no good representation as extended clauses. Thus, we present methods that detect equations between

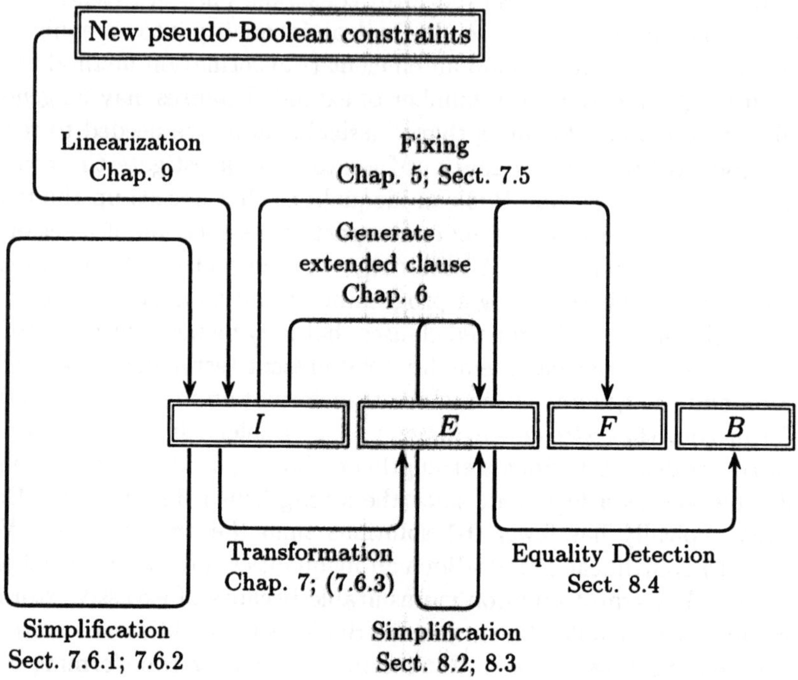

Figure 4.1 Overview of our pseudo-Boolean constraint solver

Boolean variables in Section 8.4 and also keep these equations separately. The constraint store is then of the form (I, E, F, B), where I is a set of linear pseudo-Boolean inequalities, E is a set of extended clauses, F is a set of fixed literals, and B is a set of equations between Boolean variables.

In Figure 4.1 we give an overview of the basic parts of this logic-based pseudo-Boolean constraint solver. In summary, the main goals are

- to push all information from I to E and convert E into a set of prime extended clauses.
- to detect all fixed literals, moving them to F and updating the constraint store.
- to detect all equations between Boolean variables, moving them to B and updating the constraint store.

While computation continues, we relax our requirements and just *do what can easily be done* by the following simplification techniques:

- coefficient reduction, constrained simplification and constrained transformation for linear pseudo-Boolean inequalities.
- simplifying diagonal sums, simplifying resolvents, equality detection and fixing literals for extended clauses.
- logic cut generation w.r.t. the whole constraint store by enumeration techniques, if satisfiability is not yet ensured.

The different parts of the constraint solver and the parts of the constraint store on which they depend are represented graphically in Figure 4.1. Using specific parts of the constraint store (I, E, F, B) the different techniques either reformulate constraints or derive new constraints for some other parts of the constraint store. In Figure 4.1 the techniques are listed together with the number of the section or chapter where they are presented. Each technique generates information for the part of the constraint store where it is connected with an arrow and uses information from the other connected parts.

Operationally, this logic-based constraint solver is best viewed as a collection of independently working agents that try to apply the presented techniques whenever possible. Simplification techniques are preferred and always applied in advance. If satisfiability or entailment questions cannot be answered, the logic cut generation technique of Chapter 6 is invoked and gives rise to further simplifications. This is continued as long as there are open questions. When implementing this logic-based constraint solver, the main difficulties are an efficient representation of the constraint store and a triggering of the simplification techniques such that simplification is only attempted when it seems promising. Many of the logic-based constraint processing techniques presented in this book are available in the 0-1 integer optimization package OPBDP[2].

[2] FTP: `ftp.mpi-sb.mpg.de:pub/guide/staff/barth/opbdp/opbdp.tar.Z`
WWW: `http://www.mpi-sb.mpg.de/~barth/opbdp/opbdp.html`

5

PSEUDO-BOOLEAN UNIT RESOLUTION

The traditional approach of solving a linear pseudo-Boolean optimization problem (maximize a linear pseudo-Boolean term cL subject to a set of linear pseudo-Boolean inequalities S) is to solve the linear programming relaxation of the problem and then to branch on a variable X_i having a fractional value in the current LP-solution. If there is no such variable, then the LP-optimum is the integer optimum. Branching on X_i yields the two subproblems maximize cL subject to $S, X_i = 1$ and maximize cL subject to $S, X_i = 0$, which are solved recursively by the same approach. The desired optimum is the maximum of the optima of the two subproblems. Subproblems need not be split further if either all variables are integral or if there is another subproblem having an integral solution with a greater value of the objective function.

We present an optimization method that follows a similar enumeration scheme. The main difference is that we do not consider the linear programming relaxation of the problem, but a relaxation of the problem more suitable to the logic view of a set of linear pseudo-Boolean inequalities, namely *pseudo-Boolean unit relaxation*, a generalization of the *unit relaxation* of a set of classical clauses [Hoo88a, Hoo89] to the pseudo-Boolean case. The implicit enumeration procedure can be seen as a generalization of the well-known Davis-Putnam enumeration method [DP60, Lov78] (DP) for solving the propositional satisfiability problem of a set of classical clauses (SAT).

DP-based algorithms have evolved recently into the fastest known methods for solving SAT [JW90, HHT94, Hoo93, Zha93]. This is due to proper implementation techniques and the development of good variable selection heuristics [JW90, HV94]. We present a generalization of the DP-procedure for linear pseudo-Boolean inequalities and make use of the achievements for DP. The algorithm compares well with traditional linear programming based methods on a variety of standard 0-1 integer programming benchmarks.

Pseudo-Boolean unit relaxation is the logical counterpart to the linear programming relaxation. Solving the pseudo-Boolean unit relaxation is the heart of our pseudo-Boolean constraint solver based on the logic view of pseudo-Boolean constraints. In this chapter we focus on a pure enumerative approach for solving linear pseudo-Boolean optimization problems, which corresponds to branch-and-bound in linear programming based 0-1 integer programming.

5.1 THE CLASSICAL DAVIS-PUTNAM PROCEDURE

First, we recall the classical Davis-Putnam enumeration method (DP) for solving SAT.

A classical clause $L \geq 1$ with $|L| = 1$, i.e., there is only one literal, is called a *unit clause*. The literal L_i of a unit clause is called a *unit literal*. We know that in all satisfying assignments α of a SAT-problem containing a unit clause $\{L_i\} \geq 1$, abbreviated by $L_i \geq 1$, we have $\alpha(L_i) = 1$. Given a unit literal L_i, the basic step of the DP-procedure is to replace L_i by 1 and $\overline{L_i}$ by 0 in all classical clauses followed by a simplification step. Such a step is called a *unit resolution step*. Formally, we define

$$\mathbf{ures}(L_i \geq 1, L \geq 1) := \begin{cases} L \setminus \{\overline{L_i}\} \geq 1 & \text{if } \overline{L_i} \in L, \\ \top & \text{if } L_i \in L, \\ L \geq 1 & \text{otherwise .} \end{cases} \tag{5.1}$$

Then $\mathbf{ures}(L_i \geq 1, L \geq 1)$ is the classical clause obtained from $L \geq 1$ after replacing L_i by 1 and $\overline{L_i}$ by 0 followed by the simplification step. Since we have $\alpha(L_i) = 1$ for all solutions α of a SAT-problem containing $L_i \geq 1$, we know that each solution of $\{\mathbf{ures}(L_i \geq 1, L \geq 1), L_i \geq 1\}$ is a solution of $L \geq 1$. Therefore, $L \geq 1$ is dominated by $\{\mathbf{ures}(L_i \geq 1, L \geq 1), L_i \geq 1\}$ and can be replaced by $\mathbf{ures}(L_i \geq 1, L \geq 1)$ in a SAT-problem containing $L_i \geq 1$. For a set of classical clauses S we define

$$\mathbf{ures}(L_i \geq 1, S) := \{\mathbf{ures}(L_i \geq 1, L \geq 1) \mid \begin{array}{c} L \geq 1 \in S \text{ and} \\ \mathbf{ures}(L_i \geq 1, L \geq 1) \neq \top \end{array}\}. \tag{5.2}$$

Note that there are no tautologies $\top$ in $\mathbf{ures}(L_i \geq 1, S)$. When applying **ures** to a set of classical clauses, we may generate further unit clauses and **ures** may be applicable again. *Unit resolution*, or *clausal chaining*, for a set of classical clauses S means to apply $\mathtt{ures}$ as long as there is a unit clause in S.

In the following we describe algorithms as a *transition system*, i.e., a set of *transition rules* of the form

$$\frac{\mathsf{State}_i}{\mathsf{State}_{i+1}} \text{ if } C \ , \tag{5.3}$$

which say that we replace a state State_i by a state State_{i+1} if the condition C holds. States State_i are typically tuples of the form $\langle X, Y\rangle$. A state State_n is a *normal form* of State_1 if State_n can be obtained by applying a set of transition rules to State_1 and for State_n no rule is applicable i.e., for each transition rule its condition C does not hold. In our algorithms we take care that at most one transition rule may apply to a state. Hence, if there exists a normal form, i.e., the algorithm is terminating, the normal form is unique.

Unit resolution for a set of classical clauses S is described by the following transition rule, where S is a set of classical clauses and U is a set of literals.

$$\frac{\langle S, U\rangle}{\langle \mathtt{ures}(L_i \geq 1, S), U \cup \{L_i\}\rangle} \text{ if } L_i \geq 1 \in S \tag{5.4}$$

We denote by $\langle \mathtt{ur}(S), \mathtt{ul}(S)\rangle$ the normal form obtained by applying the transition rule (5.4) as long as possible starting with $\langle S, \emptyset\rangle$. The set $\mathtt{ul}(S)$ then contains the set of all literals that occurred in a unit clause during the application of unit resolution. Moreover, $\mathtt{ur}(S)$ does not contain a unit clause and we have

$$\mathsf{Ext}(\mathtt{ur}(S) \cup \{\mathtt{ul}(S) \geq |\mathtt{ul}(S)|\}) = \mathsf{Ext}(S) \ . \tag{5.5}$$

If $\mathtt{ur}(S)$ contains the empty clause $\perp$, then S is unsatisfiable and we say that the *unit relaxation* of S is unsatisfiable. On the other hand, we know that S is satisfiable if and only if $\mathtt{ur}(S)$ is satisfiable, since each solution α of S is a solution of $\mathtt{ur}(S)$ and $\alpha(L_i) = 1$ for all $L_i \in \mathtt{ul}(S)$. Note that $\mathsf{Var}(L_i) \neq \mathsf{Var}(L_j)$ for all different L_i and L_j in $\mathtt{ul}(S)$. It is well known that the unit relaxation of S is unsatisfiable if and only if the linear programming relaxation of S is unsatisfiable [BJL86].

Given a set S of classical clauses, the DP-procedure searches a solution α of S by exploring a search tree, where each node of the tree represents a SAT-problem. Next, we define a transition system that operates on states $\langle P, Sol\rangle$. Here, P is a set of tuples $\langle S, U\rangle$, where S is a SAT-problem and U is a set of literals that have been fixed so far, representing the nodes that still need to be explored. We collect the sets of fixed literals that generate an empty SAT-problem as a set of sets of literals *Sol*.

$$\texttt{dp\_clash:}\quad \frac{\langle\{\langle S,U\rangle\}\uplus P, Sol\rangle}{\langle P, Sol\rangle}\quad \text{if } \perp\in S$$

$$\texttt{dp\_sol:}\quad \frac{\langle\{\langle S,U\rangle\}\uplus P, Sol\rangle}{\langle P, Sol\cup\{U\}\rangle}\quad \text{if } S=\emptyset$$

$$\texttt{dp\_split:}\quad \frac{\langle\{\langle S,U\rangle\}\uplus P, Sol\rangle}{\langle\{P_1,P_2\}\cup P, Sol\rangle}\quad \text{if } \begin{array}{l}\perp\notin S, S\neq\emptyset,\\ L_i=\texttt{select\_literal(S)},\\ S'=S\cup\{L_i\geq 1\},\\ P_1=\langle \texttt{ur}(S'),\texttt{ul}(S')\cup U\rangle,\\ S''=S\cup\{\{\overline{L_i}\}\geq 1\} \text{ and}\\ P_2=\langle \texttt{ur}(S''),\texttt{ul}(S'')\cup U\rangle\end{array}$$

In `dp_split` we select a so-called *branching literal* L_i occurring in some classical clause of the current SAT-problem by a procedure `select_literal`. Applying the transition system defined by the three rules `dp_clash`, `dp_sol`, and `dp_split` as long as possible on $\langle\{\langle \texttt{ur}(S),\texttt{ul}(S)\rangle\},\emptyset\rangle$, where S is a set of classical clauses, yields the normal form $\langle\emptyset, Sol\rangle$. Note that unit resolution applies at least once for each of the two subproblems introduced by `dp_split`, since we add a unit clause. Therefore, the number of literals in each of the two subproblems is smaller than the number of literals in the selected problem. Hence, the above defined transition system always terminates. We know that S is satisfiable if and only if $Sol\neq\emptyset$. From each element $U\in Sol$ we can construct a solution $\alpha\in \mathsf{Ext}(S)$ by defining $\alpha(L_i):=1$ for all $L_i\in U$ and arbitrary otherwise. The set of all solutions of S is obtained by building all solutions α for each $U\in Sol$.

For reasons of efficiency, the search tree is typically explored depth first, i.e., the set of subproblems is implemented as a "Last In, First Out"(LIFO) data structure. Since we are only interested in whether the SAT-problem is satisfiable or not, we stop as soon as the first solution is found. We define

$$\texttt{dp}(S):=\begin{cases}(\top,U) & \text{if} \quad S \text{ is satisfiable and } U\in Sol\\ (\perp,\emptyset) & \text{otherwise},\end{cases} \tag{5.6}$$

which yields $(\top,U)$ if and only if S is satisfiable and U is the first set of literals that is added to *Sol*.

5.2 DAVIS-PUTNAM FOR LINEAR PSEUDO-BOOLEAN INEQUALITIES

We obtain a DP like procedure for solving the satisfiability problem of a set of linear pseudo-Boolean inequalities by generalizing unit resolution to *pseudo-Boolean unit resolution.* The key idea of the unit resolution procedure is to fix unit literals to their only possible value and then to make obvious inferences, i.e., fixing literals and detection of (un)satisfiability. Therefore, we first need to determine whether a linear pseudo-Boolean inequality implies the fixing of a literal.

Lemma 5.2.1 *A linear pseudo-Boolean inequality $cL \geq d$ dominates $L_i \geq 1$ if and only if $c_iL_i \in cL$ and $\sum c - c_i < d$, where $\sum c$ denotes the sum over all coefficients of c.*

Proof. Let $c'L'$ be $cL \setminus \{c_iL_i\}$ and α be a solution of $cL \geq d$. Then $\alpha(c_iL_i) + \alpha(c'L') \geq d$ and $\sum c' = \sum c - c_i$.

$\Leftarrow$: Suppose that $\sum c' < d$. Because $\alpha(c'L') \leq \sum c' < d$, we must have $\alpha(c_iL_i) > 0$ and thus $\alpha(L_i) = 1$. Hence, $L_i \geq 1$ is dominated by $cL \geq d$.

$\Rightarrow$: We assume that $\sum c' \geq d$ and show that then $L_i \geq 1$ is not dominated by $cL \geq d$. Let $\alpha(L_i) = 0$ and $\alpha(L_j) = 1$ for all $L_j \in L'$, then $\alpha(cL) = \alpha(c_iL_i) + \alpha(c'L') = \sum c' \geq d$. Hence, α is a solution of $cL \geq d$, but not of $L_i \geq 1$. □

If a linear pseudo-Boolean inequality $cL \geq d$ dominates $L_i \geq 1$, we call L_i *unit literal* of $cL \geq d$.

Lemma 5.2.2 *If a linear pseudo-Boolean inequality $cL \geq d$ dominates $L_i \geq 1$ with $c_iL_i \in cL$, then $cL \geq d$ dominates $L_j \geq 1$ for all $c_jL_j \in cL$ with $c_j \geq c_i$.*

Proof. If $\sum c - c_i < d$ then $\sum c - c_j < d$, because $c_j \geq c_i$ and therefore Lemma 5.2.1 applies. But then we know that $cL \geq d$ dominates $L_j \geq 1$. □

Using Lemma 5.2.2, we see that a linear pseudo-Boolean inequality $cL \geq d$ dominates $L_i \geq 1$ for some L_i if and only if $cL \geq d$ dominates $L_1 \geq 1$, since L_1 is a literal with the largest coefficient in a linear pseudo-Boolean inequality

$cL \geq d$ in normal form. We define

$$\texttt{fixed}(cL \geq d) := \begin{cases} L_1 & \text{if } \sum c - c_1 < d \\ \bot & \text{otherwise} \end{cases} . \tag{5.7}$$

Then $\texttt{fixed}(cL \geq d) = L_1$ if and only if $cL \geq d$ dominates $L_1 \geq 1$. Moreover, if $\texttt{fixed}(cL \geq d) = \bot$, then $cL \geq d$ does not dominate any unit clause. Given a unit clause $L_i \geq 1$, we can simplify a linear pseudo-Boolean inequality $cL \geq d$ by replacing L_i by 1 and $\overline{L_i}$ by 0. Resulting constants on the left-hand side are brought to the right-hand side. Special cases arise when the linear pseudo-Boolean inequality becomes tautologous after fixing ($\top$) or unsatisfiable ($\bot$). We define

$$\texttt{fix}(L_i, cL \geq d) := \begin{cases} \top & \text{if } \ d - c_i \leq 0 \text{ and } c_i L_i \in cL, \\ cL \setminus \{c_i L_i\} \geq d - c_i & \text{if } \ d - c_i > 0 \text{ and } c_i L_i \in cL, \\ \bot & \text{if } \ \sum c - c_i < d \text{ and } c_i \overline{L_i} \in cL, \\ cL \setminus \{c_i \overline{L_i}\} \geq d & \text{if } \ \sum c - c_i \geq d \text{ and } c_i \overline{L_i} \in cL, \\ cL \geq d & \text{if } \ \text{neither } L_i \text{ nor } \overline{L_i} \text{ in } L \ . \end{cases} \tag{5.8}$$

Then $\texttt{fix}(L_i, cL \geq d)$ denotes the linear pseudo-Boolean inequality obtained by fixing a unit literal L_i in $cL \geq d$ followed by the simplification step. We have

$$\mathsf{Ext}(\{cL \geq d, L_i \geq 1\}) = \{\texttt{fix}(L_i, cL \geq d), L_i \geq 1\} \ .$$

Note that fixing a literal L_i in a linear pseudo-Boolean inequality $cL \geq d$ may produce a tautology only if $L_i \in L$ and may produce a contradiction only if $\overline{L_i} \in L$. Given a set of linear pseudo-Boolean inequalities S, we denote by $\texttt{fix}(L_i, S)$ the set of all linear pseudo-Boolean inequalities $\texttt{fix}(L_i, cL \geq d) \neq \top$ with $cL \geq d \in S$. Note that no tautologies are in $\texttt{fix}(L_i, S)$. Next, we define *pseudo-Boolean unit resolution* for a set of linear pseudo-Boolean inequalities S, the generalization of the unit resolution procedure for classical clauses to the pseudo-Boolean case.

$$\frac{\langle S, U \rangle}{\langle \texttt{fix}(L_i, S), U \cup \{L_i\} \rangle} \ \text{if } cL \geq d \in S \text{ and } L_i = \texttt{fixed}(cL \geq d) \neq \bot \tag{5.9}$$

We denote by $\langle \texttt{pbur}(S), \texttt{pbul}(S) \rangle$ the normal form obtained by applying the transition rule (5.9) as long as possible starting with $\langle S, \emptyset \rangle$. The set $\texttt{pbul}(S)$ then contains the set of all unit literals detected by $\texttt{fixed}$ during the application of pseudo-Boolean unit resolution. Moreover, $\texttt{pbur}(\mathrm{S})$ does not contain a linear pseudo-Boolean inequality $cL \geq d$ with $\texttt{fixed}(cL \geq d) \neq \bot$ and we have

$$\mathsf{Ext}(\texttt{pbur}(S) \cup \{\texttt{pbul}(S) \geq |\texttt{pbul}(S)|\}) = \mathsf{Ext}(S) \ . \tag{5.10}$$

The ***pseudo-Boolean unit relaxation*** of S is the set of all assignments α such that $\alpha(L_i) = 1$ for all $L_i \in \texttt{pbul}(U)$ if $\bot \notin \texttt{pbur}(S)$ and $\emptyset$ otherwise. If $\bot \in \texttt{pbur}(S)$, we say that the pseudo-Boolean unit relaxation of S is unsatisfiable. On the other hand, S is satisfiable if and only if $\texttt{pbur}(S)$ is satisfiable, since each solution α of S is a solution of $\texttt{pbur}(S)$ and $\alpha(L_i) = 1$ for all $L_i \in \texttt{pbul}(S)$. If S contains only classical clauses, then $\texttt{ur}(S) = \texttt{pbur}(S)$ and therefore $\bot \in \texttt{pbur}(S)$ if and only if $\bot \in \texttt{ur}(S)$. If the pseudo-Boolean unit relaxation of S is unsatisfiable, then the linear programming relaxation of S is unsatisfiable. The converse no longer holds. For example, consider

$$S = \{1 \cdot A + 1 \cdot B + 1 \cdot C \geq 2, 1 \cdot \overline{A} + 1 \cdot \overline{B} + 1 \cdot \overline{C} \geq 2\} \ . \tag{5.11}$$

Since $\texttt{fixed}(1 \cdot A + 1 \cdot B + 1 \cdot C \geq 2) = \texttt{fixed}(1 \cdot \overline{A} + 1 \cdot \overline{B} + 1 \cdot \overline{C} \geq 2) = \bot$, we have $\texttt{pbur}(S) = S$ and $\texttt{pbul}(S) = \emptyset$, thus $\bot \notin \texttt{pbur}(S)$. On the other hand, the sum of both linear pseudo-Boolean inequalities simplifies to $3 \geq 4$ and therefore the linear programming relaxation is unsatisfiable.

Next, we generalize DP to the pseudo-Boolean case. Again, we define a transition system that operates on states $\langle P, Sol\rangle$, but here P is a set of tuples $\langle S, U\rangle$, where S is a set of linear pseudo-Boolean inequalities. For the sake of completeness we include the transition system, which is almost identical to the one for classical clauses.

$$\texttt{pbdp\_clash}: \quad \frac{\langle\{\langle S,U\rangle\} \uplus P, Sol\rangle}{\langle P, Sol\rangle} \quad \text{if } \bot \in S$$

$$\texttt{pbdp\_sol}: \quad \frac{\langle\{\langle S,U\rangle\} \uplus P, Sol\rangle}{\langle P, Sol \cup \{U\}\rangle} \quad \text{if } S = \emptyset$$

$$\texttt{pbdp\_split}: \quad \frac{\langle\{\langle S,U\rangle\} \uplus P, Sol\rangle}{\langle\{P_1, P_2\} \cup P, Sol\rangle} \quad \text{if} \quad \begin{array}{l} \bot \notin S, S \neq \emptyset, \\ L_i = \texttt{select\_literal}(\mathrm{S}), \\ S' = S \cup \{L_i \geq 1\}, \\ P_1 = \langle \texttt{pbur}(S'), \texttt{pbul}(S') \cup U\rangle, \\ S'' = S \cup \{\{\overline{L_i}\} \geq 1\} \text{ and} \\ P_2 = \langle \texttt{pbur}(S''), \texttt{pbul}(S'') \cup U\rangle \end{array}$$

Only the unit resolution procedure is replaced by its generalization to the pseudo-Boolean case. Applying the transition system defined by the three rules `pbdp_clash`, `pbdp_sol`, and `pbdp_split` as long as possible starting with $\langle\{\langle\texttt{pbur(S)},\texttt{pbul(S)}\rangle\},\emptyset\rangle$, where S is a set of linear pseudo-Boolean inequalities, yields the normal form $\langle\emptyset, Sol\rangle$ and Sol represents the set of all solutions of S. We define

$$\texttt{pbdp}(S) := \begin{cases} (\top, U) & \text{if} \qquad S \text{ is satisfiable and } U \in Sol \\ (\bot, \emptyset) & \text{otherwise}, \end{cases} \tag{5.12}$$

which yields $(\top, U)$ if and only if S is satisfiable and U is the first set of literals that is added to Sol when exploring the search tree depth first and stopping after the first solution has been found.

5.3 OPTIMIZING WITH PSEUDO-BOOLEAN DAVIS-PUTNAM

Optimizing a linear pseudo-Boolean term cL subject to a set of linear pseudo-Boolean inequalities S can be done by solving a sequence of pseudo-Boolean satisfiability problems. We consider the problem of maximizing cL subject to S. Minimization works in a similar way.

The goal is to find a solution α of S such that $\alpha(cL) \geq \alpha'(cL)$ for all solutions α' of S. We will find such an assignment by solving a sequence of pseudo-Boolean satisfiability problems of the form $S_i := S \cup \{cL \geq \max_i\}$, where only $\max_i$ differs from problem to problem. We call $cL \geq \max_i$ *objective function inequality*. Suppose that $\max_0$ is such that $cL \geq \max_0$ is a tautology. Solving S_0 yields a solution α_0 of S_0 and therefore of S. A lower bound of the optimum is then $\alpha_0(cL)$. We define $\max_{i+1} := \max_i + 1$ and so exclude assignments yielding no better lower bounds than the current one. Obviously, we have $\alpha_{i+1}(cL) > \alpha_i(cL)$ for all satisfying assignments α_i of S_i and α_{i+1} of S_{i+1}. If S_i is satisfiable and S_{i+1} is unsatisfiable, then $\alpha_i(cL)$ is the desired maximum.

We can incorporate this idea into `pbdp` by replacing just the rule `pbdp_sol`. Instead of adding a computed solution that fixes the literals U, we construct a solution α such that $\alpha(L_i) = 1$ for all $L_i \in U \cup (L \backslash \overline{U})$, where $\overline{U} := \{\overline{L_i} \mid L_i \in U\}$ is the set of all negated literals in U. Then $\alpha(cL)$ is the maximal value of $\max cL$ subject to $\{U \geq |U|\}$. We add to the remaining satisfiability problems the linear pseudo-Boolean inequality $cL \geq \alpha(cL) + 1$. Thus, we ensure that for each further solution the value of the objective function is larger. If there are no remaining nodes, then the last computed $\alpha(cL)$ was optimal.

Suppose that cL is a linear pseudo-Boolean term with $c_i > 0$ for all $c_i \in c$. We define a transition system that operates on states $\langle P, \max\rangle$, where P is a set of tuples $\langle S, U\rangle$, S is a set of linear pseudo-Boolean inequalities, and U is a set of literals that are fixed so far. The current lower bound of the maximization problem is max.

$$\texttt{opbdp\_clash:} \quad \frac{\langle\{\langle S,U\rangle\}\uplus P, \max\rangle}{\langle P, \max\rangle} \quad \text{if } \bot \in S$$

$$\texttt{opbdp\_climb:} \quad \frac{\langle\{\langle S,U\rangle\}\uplus P, \max\rangle}{\langle \texttt{prop}(P, \max'+1), \max'\rangle} \quad \text{if} \quad \begin{array}{l} S = \emptyset, \\ \forall L_i \in L' \setminus \overline{U} : \alpha'(L_i) = 1 \\ \text{and } \max' = \alpha'(cL) \end{array}$$

$$\texttt{opbdp\_split:} \quad \frac{\langle\{\langle S,U\rangle\}\uplus P, \max\rangle}{\langle\{P_1, P_2\}\cup P, \max\rangle} \quad \text{if} \quad \begin{array}{l} \bot \notin S, S \neq \emptyset, \\ L_i = \texttt{select\_literal(S)}, \\ S' = S \cup \{L_i \geq 1\} \text{ and} \\ P_1 = \langle \texttt{pbur}(S'), \texttt{pbul}(S') \cup U\rangle, \\ S'' = S \cup \{\{\overline{L_i}\} \geq 1\} \text{ and} \\ P_2 = \langle \texttt{pbur}(S''), \texttt{pbul}(S'') \cup U\rangle \end{array}$$

It remains to define **prop**, which updates the remaining satisfiability problems such that only better lower bounds are generated.

$$\begin{array}{lcl} \texttt{prop}(\langle S,U\rangle \uplus P, \max) & := & \{ \begin{array}{l}\langle \texttt{pbur}(S \cup \{cL \geq \max\}), \\ \quad \texttt{pbul}(S \cup \{cL \geq \max\}) \cup U\rangle \end{array} \} \cup \texttt{prop}(P, \max) \\ \texttt{prop}(\emptyset, \max) & := & \emptyset \end{array}$$

Applying the transition system defined by the rules **opbdp_clash**, **opbdp_climb**, and **opbdp_split** as long as possible starting with $\langle\{\langle \texttt{pbur}(S), \texttt{pbul}(S)\rangle\}, -1\rangle$ yields the normal form $\langle\emptyset, \max\rangle$, where $\max \geq \alpha(cL)$ for all $\alpha \in \mathsf{Ext}(S)$ and $\max = -1$ if and only if S is unsatisfiable. We define the optimization procedure $\texttt{opbdp}(S, cL) := \max$, where max is the value computed by the above given transition system.

It is no restriction to require that cL contains only positive coefficients. Each term c_iL_i with $c_i < 0$ is equivalent to $c_i + |c_i|\overline{L_i}$, since

$$c_iL_i \;\; = \;\; c_i - c_i + c_iL_i$$

$$\begin{aligned} &= c_i - c_i \cdot (1 - L_i) \\ &= c_i - c_i\overline{L_i} \ . \end{aligned}$$

Optimizing a linear pseudo-Boolean term $\sum_{i=1}^{n} c_i L_i$ containing also negative coefficients is done by optimizing $\sum_{i=1}^{n} |c_i| L_i'$, where L_i' is L_i if $c_i > 0$ and $\overline{L_i}$ otherwise, and adding the sum over the negative coefficients c_i to the positive result of opbdp.

5.4 IMPLEMENTATION

Proper implementation and the use of appropriate data structures are important factors for the efficiency of the DP-procedure [Hoo93, HHT94, Zha93]. In a naive implementation of dp a satisfiability problem has to be copied at each node which may require excessive storage. We use the following data structure for implementing dp. Given a set of classical clauses S and the set of variables V occurring in S, we compute in advance the following index data structures. Each classical clause can be referenced by a unique identifier cid. The set of literals in a classical clause with identifier cid can be referenced by lids(cid). We can obtain a list of clause identifiers where a literal L_i occurs by cids(L_i). We represent a sub-problem of S by a list of *active variables* av, i.e., variables that are not yet fixed, and a list of *clause states*, where a clause state is a clause identifier and the number of literals nl currently active in that classical clause. Applying unit resolution for the literal L_i is done by

- deleting each element in the clause state list, where the identifier in the clause state is in cids(L_i)
- and decreasing nl by one if the identifier in the clause state is in cids($\overline{L_i}$).

We obtain a unit clause if nl is set to 1. The new unit literal is then determined by lids(cid) $\cap$ (av $\cup$ $\overline{\text{av}}$) and its variable is deleted from av. The empty clause is derived if nl is set to 0. Copying a sub-problem is now done by just copying the active clause list and the clause state list. The advantage of this data structure is that the index data structure does not change while exploring the search tree. The representation of the sub-problem in a node is very compact and detection of unit literals is straightforward.

For pbdp each linear pseudo-Boolean inequality can be referenced by a unique inequality identifier iid. Given a literal L_i, we can obtain a list of inequality identifiers where L_i occurs by $\mathsf{iids}(L_i)$. We need an additional index data structure in order to access the coefficient $c_i = \mathsf{coeff}(L_i, \mathsf{iid})$ of a given literal L_i and an inequality identifier *iid*. A sub-problem is represented by a list of active variables av and a list of inequality states. An inequality state consists of an inequality identifier iid, a current right-hand side d of the linear pseudo-Boolean inequality, and a current sum $\sum c$ over all coefficients of the active literals of the linear pseudo-Boolean inequality. For applying pseudo-Boolean unit resolution for the literal L_i we do the following for all inequality states, where the inequality identifier iid is in $\mathsf{iids}(L_i)$:

- Delete the inequality state if $\mathsf{coeff}(L_i, \mathsf{iid}) \geq d$ (eliminate tautologies).
- Otherwise decrease the current right-hand side d and the current sum over all coefficients $\sum c$ by $\mathsf{coeff}(L_i, \mathsf{iid})$.

For all inequality states, where the inequality identifier iid is in $\mathsf{iids}(\overline{L_i})$ we do the following:

- Decrease the sum over all coefficients $\sum c$ by $\mathsf{coeff}(\overline{L_i}, \mathsf{iid})$. If $\sum c$ is smaller than the current right-hand side, then the sub-problem is unsatisfiable.
- Compute the active literal $L_j \in \mathsf{av}$ for which $\mathsf{coeff}(L_j, \mathsf{iid})$ is maximal. L_j is a new unit literal, i.e., it can be fixed, if $\sum c - \mathsf{coeff}(L_j, \mathsf{iid}) < d$.

For large linear pseudo-Boolean inequalities, computing the active literal $L_j \in \mathsf{av} \cup \overline{\mathsf{av}}$ for which $\mathsf{coeff}(L_j, \mathsf{iid})$ is maximal will often yield the same literal. One improvement is to store the maximal coefficient with its literal in the inequality state and to update it only if this literal is fixed.

We implement opbdp as a slight variation of pbdp. When a satisfiable solution is found (i.e an empty inequality state list) we do not return $\top$, but calculate the maximal value max of the objective function under the current assignment. We then change destructively all copied right-hand sides d of the objective function inequality states to the new value $d - (\max + 1 - \max')$, where $\max'$ is the previous calculated maximal value. Hence, the current node becomes unsatisfiable and we backtrack until the objective function constraint is no longer violated. Then standard pbdp depth first search is continued. We follow roughly the ideas presented by Hooker [Hoo93] for incrementally solving SAT.

5.5 HEURISTICS

The selection of the branching literal [JW90, HHT94, HV94] is another important factor for the efficiency of the DP-procedure. Good literal selection heuristics can reduce the number of explored nodes by an order of magnitude. Hooker et al. [HV94] investigate several branching heuristics for the clausal case. The "Two-Sided Jeroslow-Wang" rule has been justified by a Markov chain analysis of a simplification model that selects the literal L_i when $\mathtt{pbur}(S \cup \{L_i \geq 1\})$ is the *mostly simplified* problem [HV94]. The rule says that we should branch on a variable X_i which maximizes $J(X_i) + J(\overline{X_i})$, with

$$J(L_i) := \sum_{L \geq 1 \in S : L_i \in L} 2^{-|L|} \ . \tag{5.13}$$

If $J(X_i) \geq J(\overline{X_i})$, then X_i should be selected, otherwise $\overline{X_i}$. The intention of the rule is that we branch on a variable that occurs often in short clauses and so get a small sub-problem with an increasing probability that unit literals occur.

For linear pseudo-Boolean inequalities we adapt the two-sided Jeroslow-Wang rule and branch on a variable X_i that maximizes $PBJ(X_i) + PBJ(\overline{X_i})$, with

$$PBJ(L_i) := \sum_{cL \geq d \in S : c_i L_i \in cL} \frac{c_i}{\sum c - d} \cdot 2^{-|\sum c/d|} \ . \tag{5.14}$$

If $PBJ(X_i) \geq PBJ(\overline{X_i})$, then X_i is selected, otherwise $\overline{X_i}$. We replace $|L|$, which is a measure of the length of a clause, by $|\sum c/d|$, which reduces to the length if we have classical clauses and so is a straightforward generalization. The idea is that we prefer linear pseudo-Boolean inequalities, where many of the coefficients are needed in order to satisfy the linear pseudo-Boolean inequality. We weight this preference by multiplying $2^{-|\sum c/d|}$ by $c_i/(\sum c - d)$ and so take into account the relative improvement of reaching the right-hand side when fixing the literal to 1. Alternatively, we use

$$PBJ(L_i) := \sum_{cL \geq d \in S : c_i L_i \in cL} \frac{c_i \cdot d}{(\sum c)^2} \ ,$$

which has a similar intention than (5.14) but is easier to compute and takes also very weak inequalities, i.e., where $|\sum c/d|$ is much larger than 1, into account. Further analysis is required in order to obtain a more efficient branching heuristics. For opbdp we use as alternative a greedy-like heuristics depending only on the objective function. We simply select a literal L_i with the maximal coefficient in the objective function. The idea is that a better approximation of the optimal value is obtained earlier. A better literal selection heuristics that takes into account the goal of maximizing an objective function *and* reducing the search space needs to be found.

5.6 COMPUTATIONAL RESULTS

We have implemented the procedure **opbdp** in C++ and made it available in the 0-1-optimization package OPBDP [Bar95]. First, we compare OPBDP and SATO 2.0 [Zha93], a state-of-the-art implementation of the DP-procedure, on propositional satisfiability problems in classical clausal form. We use OPBDP for solving the satisfiability problem of a set of classical clauses S by maximizing a new single variable w.r.t. S. The set of classical clauses S is satisfiable if and only if OPBDP computes the optimum 1. SATO [Zha93] uses the heuristics to branch on a literal of a shortest positive clause. We use the variant of the two-sided Jeroslow Wang heuristics, presented in Section 5.5, for OPBDP.

In Figure 5.1 we present the results running SATO and OPBDP on a variety of propositional satisfiability problems from the DIMACS test set [Tri93]. In the column "Name" we give the name of the problem. In "# V" we mention the number of variables and in "# C" the number of clauses of the problem. In "SAT" we indicate whether the problem is satisfiable or not. "Nodes" is the number of nodes explored while solving the problem and "Time" is the user cpu time used for solving the problem by OPBDP resp. SATO on a SPARC-10/31. In the average case OPBDP is about 6 times slower than SATO w.r.t. the number of nodes explored per second. Considering only hard problems, where both algorithms explore more than a million nodes, OPBDP is about 2 times slower than SATO. Taking into account that OPBDP is able to solve linear pseudo-Boolean optimizations problems, whereas SATO can just check the satisfiability of a set of classical clauses, and the use of a computationally more difficult variable selection heuristics, the implementation is reasonably efficient w.r.t. SATO.

In Figure 5.2 we present computational results obtained by solving a variety of pure 0-1 integer programming problems found in MIPLIB [BBI92] with OPBDP. In the column "Name" we give the name of the problem. In "# V" we mention the number of variables and in "# I" the number of inequalities of the problem. "Nodes" is the number of nodes explored for solving the problem and "Time" is the user cpu time used for solving the problem with OPBDP on a SPARC-10/31.

Name	# V	# C	SAT	SATO Nodes	SATO Time	OPBDP Nodes	OPBDP Time
aim-100-20-n-1	100	200	N	92442475	12368.59	32	0.05
aim-100-20-n-2	100	200	N	53797042	6196.75	1642700	363.18
aim-100-20-y1-1	100	200	Y	271346	24.75	68181	19.11
aim-100-20-y1-2	100	200	Y	32300	3.90	14723	4.69
hole6	42	133	N	770	0.05	1438	0.28
hole9	90	415	N	362880	50.15	725758	312.25
dubois20	60	160	N	2097152	155.34	4194302	546.98
dubois21	63	168	N	4194304	350.31	8388606	1011.75
ssa0432-003	435	1027	N	268848	109.09	12366	16.93
ii32b3	348	5734	Y	55	0.19	274	14.35
ii32c3	279	3272	Y	20	0.09	70	3.55
ii32e3	330	5020	Y	65	0.23	322	12.46
par-8-2-c	68	270	Y	41	0.03	20	0.06
par-8-4-c	67	266	Y	15	0.01	17	0.06
par-16-2-c	324	1292	Y	?	&	66596	185.26
par-16-4-c	349	1392	Y	2886	2.72	34495	110.43
pret60_25	60	160	N	2236416	243.67	4194302	578.00
pret60_75	60	160	N	1589784	129.20	4194302	578.76

& : aborted after 30000 seconds

Figure 5.1 Clausal Boolean Satisfiability Problems

In the "DS-JW" columns we use OPBDP with the adapted two sided Jeroslow Wang heuristics. The columns "...(opt)" describe when the optimal value is found and the columns "...(V)" describe when the optimal result is verified and OPBDP terminates. In the column "OF" we first select the literal with the largest coefficient in the objective function. Typically, this heuristics finds good or even optimal solutions quickly, whereas the verification of optimality takes too long. Hence, we only measure the time to find the optimal solution "...(opt)".

In Figure 5.3 we report the results for solving the same 0-1 integer optimization problems with CPLEX 3.0 and MINTO 1.5 [NSS94]. CPLEX is a commercial mixed integer solver based on the simplex algorithm. Integer problems are solved by a standard branch-and-bound method. In CPLEX 3.0 also cutting plane techniques are incorporated. MINTO is based on CPLEX, but additionally generates strong cuts, e.g., clique inequalities and lifted knapsack covers, at each node if possible. For CPLEX and MINTO the time when the optimal solution was found is not given. MINTO does not mention the node where the optimal solution was found. The other columns in Figure 5.3 are similar to those in Figure 5.2.

Name	#I	#V	DS-JW (opt) Nodes	Time	DS-JW (V) Nodes	Time	OF (opt) Nodes	Time
air01	46	771	2094	6.33	6302	17.68	1549	10.11
p0033	15	33	126	0.00	1083	0.05	890	0.05
p0040	23	40	215263	8.42	277520	10.81	61	0.01
p0201	133	201	51489055	18023.29	190836287	69130.40	493	0.30
bm23	20	27	6203	0.68	7165	0.79	4510	0.34
enigma	42	100	2534	0.33	2534	0.34	681	0.18
lseu	28	89	1415427	130.30	2967299	279.28	650191	59.88
misc01	70	82	214	0.06	1566	0.29	19740	2.15
misc02	52	58	114	0.01	293	0.04	141	0.06
misc03	122	159	54046	13.86	72065	20.21	425	0.13
misc07	246	259	9782421	6056.60	12037880	7552.70	164544	83.53
sentoy	30	60	312155	91.09	446968	147.09	778	0.14
stein15	36	15	9	0.00	436	0.02	6	0.01
stein27	118	27	2451	0.12	29138	1.60	17	0.01
stein45	331	45	159133	18.89	1535290	198.00	9170	1.83

& : aborted after 30000 seconds

Figure 5.2 0-1 Integer Optimization Problems from MIPLIB

Name	# I	# V	CPLEX (opt) Nodes	CPLEX (V) Nodes	Time	MINTO(V) Nodes	Time
air01	23	771	3	3	0.35	2	955.00
p0033	16	33	662	694	2.23	31	1.00
p0040	23	40	51	54	0.17	1	0.00
p0201	133	201	760	1023	25.83	841	129.00
bm23	20	27	452	452	1.88	201	27.00
enigma	21	100	22086	22086	214.68	2466	786.00
lseu	28	89	10811	15889	82.05	187	15.00
misc01	54	82	76	721	6.97	283	18.00
misc02	39	58	29	60	0.60	46	2.00
misc03	96	159	24	527	16.77	2569	378.00
misc07	212	259	2601	22832	1450.48	?	&
sentoy	30	60	595	723	4.92	162	80.00
stein15	36	15	7	85	0.38	91	1.00
stein27	118	27	8	4061	32.87	4861	125.00
stein45	331	45	29373	71595	2910.18	103207	20904.00

& : aborted after 30000 seconds

Figure 5.3 0-1 Integer Optimization Problems from MIPLIB

The presented system OPBDP compares well w.r.t. the two linear programming based systems. For the problems enigma and steinxxx we gain a least an order of magnitude. Other problems, e.g., air01 and p0201, where the optimal value of the linear programming relaxation is already near the integer optimal value, but the number of variables is high, appear to be more difficult for OPBDP. It is interesting to note that in most of the problems the optimal solution was found early when using the greedy like heuristics "OF". This indicates that better variable selections heuristics that take into account both the objective function and the reduction of the search tree need to be found. Better implementation techniques (like the one from SATO) and special handling of equality constraints are further possible ways to improve the algorithm. Note that all occurring equalities $cL = d$ have been replaced by the two inequalities $cL \leq d$ and $cL \geq d$. Therefore, the row "# I" differ from the "ROWS" column found in MIPLIB. Direct use of the equality constraints would reduce the problem size and might help to decrease the number of explored nodes.

Currently, we require that all the coefficients are integer. The results still hold when rational coefficients are allowed in the constraint inequalities and adapting the presented method is easy. This has not yet been done and therefore some problems ("p0281", "mod008", "mod010", and "pipex") have not been considered. Furthermore, we have not tried some problems with a very large number of variables. We believe that a version of OPBDP using a full repertory of logic-based constraint processing techniques could also be applied on larger problems.

5.7 PSEUDO-BOOLEAN DAVIS-PUTNAM FOR CLP

An incremental version of the pseudo-Boolean Davis-Putnam algorithm pbdp is suitable in the context of constraint logic programming, if we do not require a solved form, but global consistency. The advantage is that satisfiability is ensured at each step of computation and, as computational experience indicates, the performance is acceptable even compared to specialized 0-1 integer programming tools. Thus, a first step to a more sophisticated and efficient constraint solver for constraint logic programs over pseudo-Boolean constraints is to use incremental pbdp as constraint solver. We then satisfy at least the most important requirement imposed by constraint logic programming; global consistency. Thus, we are sure that each CLP($\mathcal{PB}$) program is executed in the same way, regardless to other improvements that may or may not speed up constraint solving.

Instead of a logic branch-and-bound algorithm, we can also use a logic branch-and-cut algorithm, which not only yields a solved form but, just as polyhedral cuts in 0-1 integer programming, may significantly reduce the number of nodes necessary to exhibit (un)satisfiability of a set of pseudo-Boolean constraints. Instead of exploring a search tree, we reformulate the current constraint store by adding valid constraints and strengthening existing constraints. The key to an efficient and powerful constraint solver is thus the development of methods for generating strong logic cuts. In the next chapter we show how to generate strong logic cuts that better approximate the set of prime extended clauses of the current constraint store by using **pbdp**.

6

LOGIC CUTS AND ENUMERATION

An unsatisfiable branch of the pseudo-Boolean Davis-Putnam enumeration algorithm `pbdp` of Chapter 5 provides information that allows us to generate a new valid extended clause that is not yet dominated by any single linear pseudo-Boolean inequality in the set of linear pseudo-Boolean inequalities to which `pbdp` is applied. Thus, we have a method for generating a logic cut for a set of linear pseudo-Boolean inequalities. This cut generation is complete, i.e., if the set of linear pseudo-Boolean inequalities is unsatisfiable, then the repeated generation of these logic cuts will detect unsatisfiability, otherwise a satisfiable branch will be hit. Thus, we have a complete pure logic cut algorithm for solving linear pseudo-Boolean constraints. Moreover, we show how to tailor the logic cut algorithm to generate logic cuts that help to decide entailment of extended clauses.

6.1 VALID EXTENDED CLAUSES BY ENUMERATION

Let us recall `pbdp` and apply it to a set of linear pseudo-Boolean inequalities S. Suppose that `pbdp_clash` applies, because we have $\perp \in S'$ in a state $\langle S', U\rangle$. For `pbdp` we delete that state, which contains useful information. Since $\perp \in S'$, we know that the set of linear pseudo-Boolean inequalities $S \cup \{U \geq |U|\}$ is unsatisfiable and that unsatisfiability is detected by pseudo-Boolean unit resolution, i.e., we have $\perp \in \mathtt{pbur}(S \cup \{U \geq |U|\})$.

From the unsatisfiability of $S \cup \{L \geq |L|\}$ we know that S dominates the negation of $L \geq |L|$. The negation of $L \geq |L|$ is $L < |L|$ and after bringing

$L < |L|$ into pseudo-Boolean normal form we obtain

$$\overline{L} \geq 1 \ , \tag{6.1}$$

where $\overline{L}$ is $\{\overline{L_i} \mid L_i \in L\}$. In summary, S dominates $\overline{L} \geq 1$, since $S \cup \{L \geq |L|\}$ is unsatisfiable. Hence, we can generate a valid extended clause w.r.t. S, whenever fixing of literals yields $\perp$. We define

$$\mathtt{gvec}(S,U) := \begin{cases} \overline{U} \geq 1 & \text{if} \quad \perp \in S \\ \top & \text{if} \quad S = \emptyset \\ \mathtt{gvec}(S',U') & \text{if} \quad \perp \notin S, S \neq \emptyset, \\ & \quad\quad L_i = \mathtt{select\_literal}(S), \\ & \quad\quad S' = \mathtt{pbur}(S \cup \{L_i \geq 1\}) \text{ and} \\ & \quad\quad U' = \mathtt{pbul}(S \cup \{L_i \geq 1\}) \cup U \ . \end{cases} \tag{6.2}$$

Operationally, we explore the search tree of **pbdp** depth first without backtracking. If a solution is found, we return $\top$, otherwise we construct a valid extended clause w.r.t. S. Hence, S dominates $\mathtt{gvec}(S, \emptyset)$. Furthermore, if $\mathtt{gvec}(S, \emptyset) = \top$, then S is satisfiable.

Let us investigate the extended clause $\overline{U} \geq 1$ returned by $\mathtt{pbdp\_gvec}(S)$ in the case that $\mathtt{pbul}(S) = \emptyset$ and $\perp \notin S$, i.e., the pseudo-Boolean unit relaxation of S is satisfiable. Its most interesting property is that it is not yet dominated by any single linear pseudo-Boolean inequality in S.

Proposition 6.1.1 *Let S be a set of linear pseudo-Boolean inequalities, where $\mathtt{pbul}(S) = \emptyset$ and $\perp \notin S$. Let $\overline{U} \geq 1 \neq \top$ be the result of $\mathtt{gvec}(S)$. Then there is no linear pseudo-Boolean inequality $cL \geq d$ in S such that $cL \geq d$ dominates $\overline{U} \geq 1$.*

Proof. Suppose that there is a linear pseudo-Boolean inequality $cL \geq d$ in S such that $cL \geq d$ dominates $\overline{U} \geq 1$. Because $\perp \notin S$ we know that $U \neq \emptyset$. Let $L_i \in U$ be the last literal that is selected by $\mathtt{select\_literal}$. We then know that $\perp \in \mathtt{pbur}(S \cup \{U \geq |U|\})$ and that $\perp \notin \mathtt{pbur}(S \cup \{U \setminus \{L_i\} \geq |U|-1\})$, since otherwise $\mathtt{gvec}(S, \emptyset) = \overline{U \setminus \{L_i\}} \geq 1$. Furthermore, we have $\overline{L_i} \notin \mathtt{pbul}(S \cup \{U \setminus \{L_i\} \geq |U| - 1\})$, since otherwise L_i cannot be selected. Let $c'L' \geq d'$ be the linear pseudo-Boolean inequality obtained from $cL \geq d$ by fixing all the literals in $\mathtt{pbul}(S \cup \{U \setminus \{L_i\} \geq |U|-1\})$ with the procedure $\mathtt{fix}$ defined by (5.8). Then $c'L' \geq d'$ is not $\perp$, since otherwise unsatisfiability would have been detected. If $c'L' \geq d'$ dominates $\overline{L_i} \geq 1$, then it is detected by pseudo-Boolean unit resolution because of Lemma 5.2.1 and thus $\overline{L_i} \in \mathtt{pbul}(S \cup \{U \setminus \{L_i\} \geq |U|-1\})$, which is a contradiction. Therefore, $c'L' \geq d'$ does not dominate $\overline{L_i} \geq 1$.

But then there is an assignment α in $\text{Ext}(c'L' \geq d')$ such that $\alpha(L_i) = 1$. Since no variable occurring in $\texttt{pbul}(S \cup \{U \setminus \{L_i\} \geq |U| - 1\})$ occurs in $c'L' \geq d'$, let $\alpha(L_j) = 1$ for all $L_j \in \texttt{pbul}(S \cup \{U \setminus \{L_i\} \geq |U| - 1\})$. Note that then $\alpha \in \text{Ext}(cL \geq d)$. Since $cL \geq d$ dominates $\overline{U} \geq 1$, we know that $\alpha \in \text{Ext}(\overline{U} \geq 1)$. But $\alpha(L_j) = 1$ for all $L_j \in U$ and thus $\alpha(\overline{U}) = 0$, which is a contradiction. Therefore, $cL \geq d$ does not dominate $\overline{U} \geq 1$. □

A straightforward way of checking whether we can derive a stronger extended clause than $\overline{U} \geq 1$ is to check whether $S \cup \{(U \setminus \{L_j\}) \cup \overline{L_j} \geq |U|\}$ is unsatisfiable for some $L_j \in U$, since then S dominates the stronger extended clause $\overline{U \setminus \{L_j\}} \geq 1$. Note that $S \cup \{U \setminus \{L_j\} \geq |L| - 1\}$ dominates $\overline{L_j} \geq 1$, since we already have derived that S dominates $\overline{U} \geq 1$. The simplest way to check unsatisfiability is to test whether $\bot \in \texttt{pbur}(S \cup \{(U \setminus \{L_j\}) \cup \overline{L_j} \geq |U|\})$. Such a test requires at most $|U|$ times the application of pbur. Since pbur can be efficiently implemented, as shown in Section 5.4, such a test is suitable. Whenever we have eliminated a literal L_j, we can still try to eliminate not yet tried literals by the same method. If the generated extended clause is still too weak, we can also apply a restricted version of pbdp for the unsatisfiability test. For example, we can limit the depth of the search tree to a fixed bound. Note that if exploring this search tree yields a solution, we have also shown satisfiability of the current constraint store.

6.2 A PURE LOGIC CUT ALGORITHM

By Proposition 6.1.1 the procedure pbdp_gvec generates a logic cut not yet dominated by any single linear pseudo-Boolean inequality in S or detects satisfiability of a set of pseudo-Boolean inequalities S. Since only variables from S occur in the logic cut, there is a finite number of not yet dominated logic cuts. Thus, repeatedly adding the result of gvec to S and applying gvec again will finally either generate $\top$, indicating that S is satisfiable, or the pseudo-Boolean unit relaxation will become unsatisfiable. We define

$$\texttt{plc}(S, F) := \begin{cases} \{\bot\} & \text{if } \bot \in S \\ S \cup \{F \geq |F|\} & \text{if } \bot \notin S \text{ and } \texttt{gvec}(S) = \top \\ \texttt{plc}(S', F') & \text{if } \bot \notin S, \\ & \quad \texttt{gvec}(S, \emptyset) = \overline{U} \geq 1 \neq \top, \\ & \quad S' = \texttt{pbur}(S \cup \{\overline{U} \geq 1\}) \text{ and} \\ & \quad F' = \texttt{pbul}(S \cup \{\overline{U} \geq 1\}) \cup F \ ; \end{cases} \tag{6.3}$$

a pure logic cut algorithm for solving linear pseudo-Boolean inequalities.

Theorem 6.2.1 *Let* $S' = \texttt{plc}(\texttt{pbur}(S), \texttt{pbul}(S))$*, where* S *is a set of linear pseudo-Boolean inequalities. Then* $\mathsf{Ext}(S) = \mathsf{Ext}(S')$ *and* $S' = \{\bot\}$ *if* S *is unsatisfiable.*

Proof. By definition (6.3) of `plc`, we know that `gvec` is only applied on a set of linear pseudo-Boolean inequalities S'' with $\texttt{pbul}(S'') = \emptyset$. By Proposition 6.1.1 we know that each clause generated by `gvec` is valid w.r.t. S. Since `plc` repeatedly adds valid extended clauses, we have $\mathsf{Ext}(S) = \mathsf{Ext}(S')$. Furthermore, we know that `gvec` always generates $\top$, or a clause consisting of variables from S that is not yet dominated by the current set of linear pseudo-Boolean inequalities. Since there is only a finite number of these clauses, either $\top$ is generated, or the pseudo-Boolean unit relaxation of the current problem is unsatisfiable. If $\top$ is generated, then the current equivalent problem is returned and the problem is known to be satisfiable. If the pseudo-Boolean unit relaxation of the current problem is unsatisfiable, then $\bot$ is in the current problem and $\{\bot\}$ is returned. □

We say that $S' = \texttt{plc}(\texttt{pbur}(S), \texttt{pbul}(S))$ is closer to the solved form than S. Note that if S dominates an extended clause $L \geq d$, then there is a single extended clause in the solved form of S, the set of all prime extended clauses of S, dominating $L \geq d$. Obviously, there are more extended clauses that are dominated by an extended clause in S' than in S, if at least one clause generated by `gvec` is added.

The pure logic cut algorithm `plc` always adds a new extended clause. Despite the advantage that `plc` produces a stronger reformulation of the current problem, and therefore is suitable in the context of constraint logic programming, one may argue that `pbdp` would always be faster in deciding satisfiability than `plc`. We think that `plc` has two properties that allow for a more efficient handling of the pseudo-Boolean satisfiability problem than `pbdp`.

- Adding strong logic cuts changes the problem and so the variable selection heuristics may decide to try other branches. In other words, we do not completely explore a subtree if we choose the wrong literal early.
- We can simplify the current problem after the generation of a strong logic cut, which may change the problem completely. In particular, if the problem is unsatisfiable it is likely that simplification techniques apply and the

problem is solved much faster. Since the reformulation of the problem is closer to the set of its prime extended clauses, again the variable selection heuristics may try out completely different branches that are more promising.

In Chapter 8 we discuss several simplification techniques working on extended clauses and in Section 7.6 simplification techniques for linear pseudo-Boolean inequalities. Suppose that these simplification techniques are realized by a procedure `simplify`, then we obtain an enhanced pure logic cut algorithm by redefining

$$S' = \texttt{simplify}(S \cup \{\overline{U} \geq 1\})$$

in (6.3). The efficiency of a pure logic cut constraint solver depends heavily on the simplification techniques and the logic cuts that these simplification techniques generate. In an implementation we have to provide an incremental version of the simplification methods that are triggered by the newly added logic cut generated by **gvec**, where we assume that the set of linear pseudo-Boolean inequalities is initially in fully simplified form.

Example 6.2.2 We apply `plc` on $S :=$

$$\{A{+}F \geq 1, \overline{A}{+}B{+}\overline{C} \geq 1, \overline{A}{+}\overline{B}{+}D{+}E \geq 1, \overline{A}{+}\overline{C}{+}\overline{D} \geq 1, \overline{A}{+}\overline{B}{+}D{+}\overline{E} \geq 1\} \ .$$

We start fixing $A = 1$ and obtain $\textbf{pbur}(S') := \textbf{pbur}(S \cup \{A \geq 1\}) =$

$$\{B + \overline{C} \geq 1, \overline{B} + D + E \geq 1, \overline{C} + \overline{D} \geq 1, \overline{B} + D + \overline{E} \geq 1\} \ .$$

Next, we select $B = 1$ and get $\textbf{pbur}(S'') := \textbf{pbur}(\textbf{pbur}(S') \cup \{B \geq 1\}) =$

$$\{D + E \geq 1, \overline{C} + \overline{D} \geq 1, D + \overline{E} \geq 1\} \ .$$

By selecting $C = 1$ we get a contradiction, since we derive as fixed literals first $\overline{D} = 1$ from the second clause in $\textbf{pbur}(S'')$ and then the first clause implies $E = 1$, whereas the third implies $\overline{E} = 1$. Thus, we know that

$$\overline{A} + \overline{B} + \overline{C} \geq 1$$

is dominated by S. Let us check whether all selections are necessary for deriving the contradiction. Indeed, we see that already $\bot \in \textbf{pbur}(S \cup \{A{+}C \geq 2\})$, since the second extended clause of S forces $B = 1$. Thus, we derive the stronger extended clause

$$\overline{A} + \overline{C} \geq 1 \ .$$

Next, we add $\overline{A} + \overline{C} \geq 1$ to S and delete dominated extended clauses. We obtain $\hat{S} :=$

$$\{A + F \geq 1, \overline{A} + \overline{C} \geq 1, \overline{A} + \overline{B} + D + E \geq 1, \overline{A} + \overline{B} + D + \overline{E} \geq 1\} .$$

We restart solving $\hat{S}$ by again selecting first $A = 1$ and obtain $\texttt{pbur}(\hat{S}') :=$

$$\{\overline{B} + D + E \geq 1, \overline{B} + D + \overline{E} \geq 1\} .$$

Note that $\overline{C} = 1$ is now already detected. We have a stronger formulation and pseudo-Boolean unit resolution can detect more fixed literals. We again may select B, but after selecting D we obtain $\emptyset$ and thus S is satisfiable. Hence, we replace S by $\hat{S}$ in the current constraint store.

Suppose that our variable selection heuristics selects $A = B = \overline{D} = 1$ for $\hat{S}$ and thus again derives $\perp$. Then we obtain the valid extended clause

$$\overline{A} + \overline{B} + D \geq 1$$

and adding $\overline{A} + \overline{B} + D \geq 1$ to $\hat{S}$ yields, after deleting dominated extended clauses, $\widetilde{S} :=$

$$\{A + F \geq 1, \overline{A} + \overline{C} \geq 1, \overline{A} + \overline{B} + D \geq 1\} .$$

Now for $\widetilde{S}$ any variable selection heuristics in $\texttt{gvec}$ derives satisfiability. We again have a stronger formulation of S and replace $\hat{S}$ by $\widetilde{S}$ in the current constraint store.

6.3 ENTAILMENT AND LOGIC CUTS

Deciding entailment of pseudo-Boolean constraints is a basic functionality in the context of concurrent constraint logic programming [JM94]. It is sufficient to decide entailment of extended clauses, since arbitrary entailment questions can be reformulated by the linearization technique of Chapter 9 and transformation technique of Chapter 7 to a set of entailment questions on extended clauses. Let S be the set of linear pseudo-Boolean inequalities representing the current constraint store. The problem is to check whether an extended clause $L \geq d$ is dominated by S.

We know that S dominates $L \geq d$ if and only if S together with the negation of $L \geq d$, i.e., $\overline{L} \geq |L| + 1 - d$, is unsatisfiable. A straightforward way for checking entailment of $L \geq d$ is therefore $\texttt{pbdp}$. We know that S dominates $L \geq d$ if and only if

$$\texttt{pbdp}(S \cup \{\overline{L} \geq |L| + 1 - d\}) = (\perp, \emptyset) .$$

Checking entailment by that method has the disadvantage that we have to explore a search tree for each entailment question and that after having decided domination, we have not gained any new information and the constraint store is unchanged. Therefore, we introduce a logic cut generation method that successively generates new logic cuts until entailment becomes easily decidable.

First, we restrict ourselves to the case $d = 1$, i.e., deciding whether or not a classical clause $L \geq 1$ is dominated by S. We know immediately that $L \geq 1$ is dominated by S if there is an extended clause in S dominating $L \geq 1$. If there is no such extended clause we try to generate it with our logic cut generation method based on the enumeration technique of Chapter 5. Again, the basic idea is to check unsatisfiability of the negation of $L \geq 1$ and S. The negation of $L \geq 1$ is $\overline{L} \geq |L|$, i.e., all literals in L are fixed to 0. Thus, we first apply pseudo-Boolean unit resolution and calculate

$$S' = \mathbf{pbur}(S \cup \{\overline{L} \geq |L|\}) \ .$$

If $\bot \in S'$, we know that $L \geq 1$ is dominated. If S' is empty, we know that $S \cup \{\overline{L} \geq |L|\}$ is satisfiable and thus S does not dominate $L \geq 1$. The remaining case is that S' still contains linear pseudo-Boolean inequalities. We then apply **gvec** on S'. If $\mathbf{gvec}(\mathbf{pbur}(S'), \mathbf{pbul}(S'))$ is $\top$, then $L \geq 1$ is not dominated, since S', and thus $S \cup \{\overline{L} \geq |L|\}$, is satisfiable. The other possibility is that **gvec** generates a clause $L' \geq 1$ dominated by S'. But our goal is to generate a valid extended clause dominated by S. Since we know that $S \cup \{\overline{L} \geq |L|, \overline{L'} \geq |L'|\}$ is unsatisfiable, we know that $S \cup \{\overline{L \cup L'} \geq |L| + |L'|\}$ is unsatisfiable. Thus, we obtain a valid extended clause dominated by S, namely $L \cup L' \geq 1$, which is a logic cut of S. By a repeated application of that procedure, we add not yet dominated logic cuts of the form $L \cup L' \geq 1$, where L' varies and L is fixed. Since there is only a finite number of these logic cuts, again, we are sure that either satisfiability of $S \cup \{\overline{L} \geq |L|\}$ is detected, and thus $L \geq 1$ is not dominated, or unsatisfiability of $S \cup \{\overline{L} \geq |L|\}$ is detected, and thus $L \geq 1$ is dominated. We restart the enumeration each time after adding a generated clause, since simplification may lead to a stronger formulation of the current constraint store and thus deciding entailment becomes more efficient. We get the following algorithm deciding entailment of a classical clause $L \geq 1$ w.r.t. S.

Algorithm 6.3.1 *Entailment of a classical clause*

$\mathtt{entail\_classclause}(S, L \geq 1)$
  if $\exists L' \geq d' \in S : L' \geq d' \models L \geq 1$ **then return** $(S, \top)$ **endif**
  $S' := \mathbf{pbur}(S, \overline{L} \geq |L|)$

```
        if ⊥ ∈ S' then return (simplify(S ∪ {L ≥ 1}), ⊤)
        elseif S' = ∅ then return (S, ⊥)
        endif
        R := gvec(pbur(S'), pbul(S'))
        if R = ⊤ then return (S, ⊥)
        elseif R = L' ≥ 1 then
                S'' := simplify(S ∪ {L ∪ L' ≥ 1})
                return entail_classclause(S'', L ≥ 1)
        endif
end entail_classclause
```

We know that entail_classclause$(S, L \geq 1)$ returns $(S', \top)$ if and only if S dominates $L \geq 1$ and it returns $(S', \perp)$ if and only if S does not dominate $L \geq 1$. In both cases we have $\mathsf{Ext}(S') = \mathsf{Ext}(S)$ and S' is stronger than S. We replace the current constraint store S by S'. Thus, Algorithm 6.3.1 decides entailment of a classical clause and strengthens the current constraint store.

When deciding entailment of an extended clause $L \geq d$ with $d > 1$, we apply in principle the same algorithm. First, we check whether $L \geq d$ is dominated by an extended clause in S. If yes, it is entailed. Otherwise we choose a strict reduction $L' \geq 1$ of $L \geq d$ with $L' \subseteq L$ and $|L \setminus L'| = d - 1$. Additionally, we require that $L' \geq 1$ is not yet dominated by an extended clause in S. If all reductions $L' \geq 1$ of $L \geq d$ are dominated by an extended clause in S, then $L \geq d$ is dominated by S, since the set of all reductions $L' \geq 1$ is equivalent to $L \geq d$, see Theorem 7.1.5, and we have decided entailment. Next, we check whether $L' \geq 1$ is dominated by S with entail_classclause. If this is not the case, we know that $L \geq d$ is not dominated by S. If $L' \geq 1$ is dominated, then the constraint store is modified and there exists an extend clause in the current constraint store dominating $L' \geq 1$. Next, we again select a strict reduction of $L \geq d$, which cannot be $L' \geq 1$, and iterate the whole process. Since there are only finitely many reductions of $L \geq d$, the iteration must terminate. Hence, we have a decision procedure for deciding entailment of an extended clause $L \geq d$.

Algorithm 6.3.2 *Entailment of an extended clause*

```
entail_extendedclause(S, L ≥ d)
        if ∃L̂ ≥ d̂ ∈ S : L̂ ≥ d̂ ⊨ L ≥ d then return (S, ⊤) endif
        if ∃L' ≥ 1 ∈ SRed(L ≥ d) ∧ ∀L̂ ≥ d̂ ∈ S : L̂ ≥ d̂ ⊭ L' ≥ 1 then
                if entail_classclause(S, L' ≥ 1) = (S', ⊥) then
                        return (S', ⊥)
```

```
            elseif entail_classclause(S, L' ≥ 1) = (S', ⊤) then
                  return entail_extendedclause(S', L ≥ d)
            endif
      else
            return (simplify(S ∪ {L ≥ d}), ⊤)
      endif
end entail_extendedclause
```

At a first sight it seems that Algorithm 6.3.2 would not be very efficient, since in the extreme case all strict reductions have to be checked for entailment. Again, the key point is that simplification takes place. In Section 8.2 we present an efficient method that generates extended clauses with right-hand side greater than 1. So, for example, when adding a classical clause $L' \geq 1$ it is likely that an extended clause $L' \cup \{L_j\} \geq 2$ is generated, which dominates $L' \geq 1$. Or even $L' \cup \{L_j, L_k\} \geq 3$ and so on. Thus, we hope that the simplification techniques of Section 8.2 will generate an extended clause dominating $L \geq d$ after only a few classical clauses have been generated.

The most important advantage of the presented entailment procedure is that not only entailment is decided, but the current constraint store is reformulated and gets closer to the solved form, the set of all prime extended clauses. Hence, the constraint solver is in a state where current questions about entailment can be answered, but the set of all prime extended clauses is not necessarily generated. Therefore, we spend computation time to answer questions and not to provide a solved form when unnecessary.

7

LINEAR PSEUDO-BOOLEAN INEQUALITIES AND EXTENDED CLAUSES

Transforming linear pseudo-Boolean inequalities into an equivalent set of extended clauses allows us to handle linear pseudo-Boolean constraints with specialized methods for extended clauses [Bar93c, Bar94]. Furthermore, deciding whether a linear pseudo-Boolean inequality is dominated by a set of pseudo-Boolean constraints reduces to the simpler problem of deciding whether extended clauses are dominated; an issue for which our solved form of pseudo-Boolean constraints is well-suited.

The idea of reformulating linear pseudo-Boolean inequalities in a simpler class of formulas without changing the solution space goes back to Hammer and Granot [HR68, GH71]. They showed how to express linear pseudo-Boolean inequalities as an equivalent set of classical clauses. Their attempt failed mainly because of the large number of generated classical clauses [GG80]. We express a linear pseudo-Boolean inequality as an equivalent set of extended clauses. Typically, substantially fewer extended clauses than classical clauses are needed to represent a linear pseudo-Boolean inequality.

In this chapter we present a transformation method efficiently producing prime extended clauses of a linear pseudo-Boolean inequality. We adapt cutting plane techniques, known from 0-1 integer programming, for generating valid extended clauses w.r.t. a linear pseudo-Boolean inequality in Section 7.1 and prove the equivalence of the set of generated extended clauses and the linear pseudo-Boolean inequality. Specialized strong redundancy criteria, that ensure that the minimal number of needed extended clauses is generated and that allow for an efficient redundancy test, are presented in Section 7.2. We sketch an implementation of the transformation method in Section 7.3. An important improvement is the use of symmetries along with a compact representation of linear pseudo-Boolean inequalities and extended clauses, introduced in Section 7.4. We use pseudo-Boolean unit resolution, introduced in Chapter 5, in order to detect some of the fixed literals of a set of linear pseudo-Boolean inequalities. In the context of constraint logic programming, we typically already have a set of pseudo-Boolean constraints in the constraint store. Hence, we need not generate extended clauses that are already dominated by the constraint store. *Constrained transformation*, presented in Section 7.6, takes the information of the constraint store into account and thus we may reduce the number of generated extended clauses. In this context we also present simplification methods that strengthen a linear pseudo-Boolean inequality w.r.t. the constraint store and introduce methods that only partially transform a linear pseudo-Boolean inequality.

Besides the use inside our constraint solver for CLP($\mathcal{PB}$), we also relate the produced set of extended clauses to extended cover inequalities [NW88]. We show in Section 7.7 that the method generates the set of all *strongest, non-redundant* extended cover inequalities w.r.t. a linear pseudo-Boolean inequality, which are used as cutting planes in traditional 0-1 integer programming [NW88]. We apply the presented technique in a preprocessing phase for 0-1 integer programming problems and give some encouraging computational results. Another application is the implementation of one of the basic deduction rules of *Generalized Resolution* [Hoo92a], introduced in Section 7.8, that allows us to apply several resolution steps at once.

7.1 GENERATING VALID EXTENDED CLAUSES

We show how to generate a set of extended clauses equivalent to a linear pseudo-Boolean inequality. We define the *strongest extended clause* of a linear pseudo-Boolean inequality, and relate it to *cutting plane* inequalities known from 0-1 integer programming [NW88]. The equivalence of the set of all strongest extended clauses of all strict reductions of a linear pseudo-Boolean inequality and the linear pseudo-Boolean inequality itself is the main theorem of this section.

7.1.1 The Strongest Extended Clause

We already have seen in Section 3.3.2 that linear combination and integer rounding are well-known techniques of 0-1 integer programming for deriving valid inequalities [NW88]. We adopt these techniques in order to generate valid extended clauses.

Let us divide a linear pseudo-Boolean inequality in normal form

$$c_1 L_1 + \cdots + c_n L_n \geq d \tag{7.1}$$

by a positive integer k and round the fractional coefficients of the left-hand side yielding

$$\lceil c_1/k \rceil L_1 + \cdots + \lceil c_n/k \rceil L_n \geq d/k$$

with $\lceil a \rceil := \min(\{b \in \mathbb{N} \mid b \geq a\})$. Because rounding only increases the left-hand side, we have generated a valid inequality w.r.t. (7.1). Since the coefficients $\lceil c_i/k \rceil$ are integer, the left-hand side is integer for all assignments. Therefore, the right-hand side d/k can be replaced by the largest integer less than or equal to d/k, which is $\lceil d/k \rceil$. We obtain the valid linear pseudo-Boolean inequalities

$$\lceil c_1/k \rceil L_1 + \cdots + \lceil c_n/k \rceil L_n \geq \lceil d/k \rceil \ ,$$

dominated by (7.1), for all positive integers k. These linear pseudo-Boolean are a subclass of the so-called *Chvátal-Gomory cutting plane* inequalities [Gom58, NW88], obtained from a linear combination of the linear pseudo-Boolean inequality and the valid bounds $L_i \geq 0$ and $-L_i \geq -1$ for the literals L_i followed by integer rounding of the coefficients and the right-hand side. We obtain the valid extended clause

$$L_1 + \cdots + L_n \geq \lceil d/c_1 \rceil \ , \tag{7.2}$$

dominated by (7.1), if we set $k = c_1$, since then the $\lceil c_i/k \rceil$ are all 1. The extended clause (7.2) derived from a linear pseudo-Boolean inequality $cL \geq d$ is denoted by $\mathsf{CP}(cL \geq d)$ and is dominated by $cL \geq d$.

Example 7.1.1 Let $cL \geq d$ be

$$5 \cdot L_1 + 4 \cdot L_2 + 3 \cdot L_3 + 2 \cdot L_4 + 1 \cdot L_5 \geq 11$$

then $\mathsf{CP}(cL \geq d)$ is

$$L_1 + L_2 + L_3 + L_4 + L_5 \geq 3 \ .$$

For a set of linear pseudo-Boolean inequalities S we denote by $\mathsf{CP}(S)$ the set of extended clauses $\mathsf{CP}(cL \geq d)$ for all $cL \geq d$ in S. The cutting plane operation CP generates a valid extended clause $L \geq d'$ dominated by the linear pseudo-Boolean inequality $cL \geq d$. We are now interested in the greatest integer β such that $L \geq \beta$ is dominated by $cL \geq d$. We call $L \geq \beta$ *strongest extended clause* of $cL \geq d$.

Proposition 7.1.2 *The strongest extended clause of a linear pseudo-Boolean inequality $cL \geq d$ is*

$$L \geq \beta \ , \tag{7.3}$$

where β is the smallest β such that $\sum_{i=1}^{\beta} c_i \geq d$, determined by the condition

$$\sum_{i=1}^{\beta-1} c_i < d \leq \sum_{i=1}^{\beta} c_i \ . \tag{7.4}$$

The extended clauses $L \geq \beta + l$ are not dominated by $cL \geq d$ for all $l \geq 1$.

Proof.

[a] First, we show that $cL \geq d$ dominates $L \geq \beta$. Because $cL \geq d$ is in pseudo-Boolean normal form, we know that $c_1 \geq \cdots \geq c_n$. Thus, for all $1 \leq k \leq n$ we have $\sum_{i=1}^{k} c_i \geq \sum_{i \in K} c_i$ for all $K \subseteq \{1, \ldots, n\}$ with $|K| = k$. Let k be $\beta - 1$. Since $\sum_{i=1}^{\beta-1} c_i < d$, we know that $\sum_{i \in K} c_i < d$ for all $K \subseteq \{1, \ldots, n\}$ with $|K| = \beta - 1$. Consequently, for all solutions $\alpha \in \mathsf{Ext}(cL \geq d)$ we have $\alpha(L) > \beta - 1$ and therefore $\alpha(L) \geq \beta$.

[b] Next, we show that $cL \geq d$ does not dominate $L \geq \beta + l$ for all $l \geq 1$. Let α be an assignment such that $\alpha(L_1) = \ldots = \alpha(L_\beta) = 1$ and $\alpha(L_{\beta+1}) = \ldots = \alpha(L_n) = 0$. Then $\alpha \in \mathsf{Ext}(cL \geq d)$ because $\sum_{i=1}^{\beta} c_i \geq d$, but $\alpha \notin \mathsf{Ext}(L \geq \beta + l)$ for all $l \geq 1$ because $\alpha(L) = \beta$. □

For a linear pseudo-Boolean inequality $cL \geq d$ we denote by $\mathsf{SCP}(cL \geq d)$ the strongest extended clause $L \geq \beta$ of Proposition 7.1.2. For a set of linear pseudo-Boolean inequalities S we denote by $\mathsf{SCP}(S)$ the set of all extended clauses $\mathsf{SCP}(cL \geq d)$ with $cL \geq d$ in S. From Proposition 7.1.2 we know that the degree of $\mathsf{SCP}(cL \geq d)$ is greater than or equal to the degree of $\mathsf{CP}(cL \geq d)$; therefore $\mathsf{SCP}(cL \geq d)$ dominates $\mathsf{CP}(cL \geq d)$.

Example 7.1.3 Let $cL \geq d$ again be

$$5 \cdot L_1 + 4 \cdot L_2 + 3 \cdot L_3 + 2 \cdot L_4 + 1 \cdot L_5 \geq 11$$

as in Example 7.1.1, then

$$\mathsf{CP}(cL \geq d) = \mathsf{SCP}(cL \geq d) = L_1 + L_2 + L_3 + L_4 + L_5 \geq 3 \ .$$

If $cL \geq d$ is

$$5 \cdot L_1 + 4 \cdot L_2 + 3 \cdot L_3 + 2 \cdot L_4 + 1 \cdot L_5 \geq 10 \ ,$$

then

$$\mathsf{CP}(cL \geq d) \quad = \quad L_1 + L_2 + L_3 + L_4 + L_5 \geq 2$$

and

$$\mathsf{SCP}(cL \geq d) \quad = \quad L_1 + L_2 + L_3 + L_4 + L_5 \geq 3 \ .$$

Note that $\mathsf{SCP}(cL \geq d)$ strictly dominates $\mathsf{CP}(cL \geq d)$.

7.1.2 Strict Reductions

We know that all reductions of a linear pseudo-Boolean inequality $cL \geq d$ are valid w.r.t. $cL \geq d$. Hence, for all strict reductions $c'L' \geq d'$ of $cL \geq d$ we know that $\mathsf{SCP}(c'L' \geq d')$ is a valid extended clause w.r.t. $cL \geq d$.

Example 7.1.4 A strict reduction of

$$5 \cdot L_1 + 4 \cdot L_2 + 3 \cdot L_3 + 2 \cdot L_4 + 1 \cdot L_5 \geq 11$$

is

$$5 \cdot L_1 + 4 \cdot L_2 + 3 \cdot L_3 + 2 \cdot L_4 \geq 10$$

obtained by eliminating L_5. Its strongest extended clause is

$$L_1 + L_2 + L_3 + L_4 \geq 3 \ .$$

We now show that $\mathsf{SCP}(\mathsf{SRed}(cL \geq d))$, the set of strongest extended clauses derived from all strict reductions of a linear pseudo-Boolean inequality $cL \geq d$, is equivalent to $cL \geq d$.

Theorem 7.1.5 *Let $cL \geq d$ be an arbitrary linear pseudo-Boolean inequality, then*

$$\mathsf{Ext}(cL \geq d) = \mathsf{Ext}(\mathsf{SCP}(\mathsf{SRed}(cL \geq d))) \ . \tag{7.5}$$

Proof.

[a] $\mathsf{Ext}(cL \geq d) \subseteq \mathsf{Ext}(\mathsf{SCP}(\mathsf{SRed}(cL \geq d)))$:
For all strict reductions $c'L' \geq d'$ of $cL \geq d$ we know that $\mathsf{Ext}(cL \geq d) \subseteq \mathsf{Ext}(c'L' \geq d')$ because $cL \geq d$ dominates $c'L' \geq d'$. Hence,

$$\mathsf{Ext}(cL \geq d) \subseteq \bigcap_{c'L' \geq d' \in \mathsf{SRed}(cL \geq d)} \mathsf{Ext}(c'L' \geq d') \quad = \quad \mathsf{Ext}(\mathsf{SRed}(cL \geq d)) \ .$$

It remains to show $\mathsf{Ext}(\mathsf{SRed}(cL \geq d)) \subseteq \mathsf{Ext}(\mathsf{SCP}(\mathsf{SRed}(cL \geq d)))$. Since a strongest extended clause of a linear pseudo-Boolean inequality is valid w.r.t. this inequality, we have $\mathsf{Ext}(c'L' \geq d') \subseteq \mathsf{Ext}(\mathsf{SCP}(c'L' \geq d'))$ for all $c'L' \geq d'$ of $cL \geq d$. Because the intersection of sets T_i is a subset of the intersection of sets T_i' if $T_i \subseteq T_i'$, we have $\mathsf{Ext}(\mathsf{SRed}(cL \geq d)) \subseteq \mathsf{Ext}(\mathsf{SCP}(\mathsf{SRed}(cL \geq d)))$.

[b] $\mathsf{Ext}(\mathsf{SCP}(\mathsf{SRed}(cL \geq d))) \subseteq \mathsf{Ext}(cL \geq d)$:
We show that $\alpha \notin \mathsf{Ext}(\mathsf{SCP}(\mathsf{SRed}(cL \geq d)))$ if $\alpha \notin \mathsf{Ext}(cL \geq d)$ from which the theorem follows. Suppose that $\alpha \notin \mathsf{Ext}(cL \geq d)$. We show that in this case there is a strict reduction of $cL \geq d$ such that α is not a solution of the strongest extended clause of this reduction, and therefore $\alpha \notin \mathsf{Ext}(\mathsf{SCP}(\mathsf{SRed}(cL \geq d)))$. Since $\alpha \notin \mathsf{Ext}(cL \geq d)$, we know that $\alpha(cL) = d' < d$. Let Y be the set of literals $\{L_i \mid \alpha(L_i) = 1\}$ and $Z = L \backslash Y$, then $\sum_{i:L_i \in Y} c_i = d' < d$. Let $c'L' \geq d'$ be $\sum_{i:L_i \in Z} c_i L_i \geq d - d'$. Because $d' < d$, we know that $d - d' \geq 1$, and therefore $c'L' \geq d'$ is a strict reduction of $cL \geq d$, obtained by eliminating the literals that are mapped to zero by α. Let $\mathsf{SCP}(c'L' \geq d')$ be $L' \geq \beta$. We know that $\beta \geq 1$ because $d - d' \geq 1$. Because $\alpha(L') = 0$, we derive that $\alpha \notin \mathsf{Ext}(\mathsf{SCP}(c'L' \geq d'))$, and therefore $\alpha \notin \mathsf{Ext}(\mathsf{SCP}(\mathsf{SRed}(cL \geq d)))$, which proves the theorem. □

Theorem 7.1.5 suggests a simple procedure for transforming a linear pseudo-Boolean inequality into an equivalent set of extended clauses. Unfortunately, a large number of extended clauses may be produced. Obviously, we preserve equivalence if we consider only extended clauses that are not dominated by others.

Example 7.1.6 The set of all strongest extended clauses of all strict reductions of

$$5 \cdot L_1 + 4 \cdot L_2 + 3 \cdot L_3 + 2 \cdot L_4 + 1 \cdot L_5 \geq 11 \tag{7.6}$$

is

$L_1 + L_2 + L_3 + L_4 + L_5 \geq 3$	$L_1 + L_2 + L_3 \geq 2$	$L_1 + L_2 + L_3 + L_5 \geq 2$	
$L_1 + L_2 + L_4 + L_5 \geq 2$	$L_1 + L_3 + L_4 \geq 2$	$L_2 + L_3 + L_4 + L_5 \geq 2$	
$L_1 + L_2 + L_3 + L_4 \geq 3$	$L_1 + L_2 + L_4 \geq 2$	$L_1 + L_2 + L_5 \geq 2$	
$L_1 + L_3 + L_4 + L_5 \geq 2$	$L_1 + L_3 + L_5 \geq 1$	$L_1 + L_4 + L_5 \geq 1$	
$L_2 + L_3 + L_4 \geq 2$	$L_2 + L_3 + L_5 \geq 1$	$L_2 + L_4 + L_5 \geq 1$	
$L_3 + L_4 + L_5 \geq 1$	$L_1 + L_2 \geq 1$	$L_1 + L_3 \geq 1$	
$L_1 + L_4 \geq 1$	$L_1 + L_5 \geq 1$	$L_2 + L_3 \geq 1$	
$L_2 + L_4 \geq 1$	$L_2 + L_5 \geq 1$	$L_3 + L_4 \geq 1$	$L_1 \geq 1$.

Note that there are 25 strict reductions and therefore 25 extended clauses. If we delete all extended clauses that are dominated by others, we obtain

$$L_1 + L_2 + L_3 + L_4 \geq 3 \qquad L_1 + L_2 + L_5 \geq 2 \qquad L_1 \geq 1 \ .$$

The set of these 3 extended clauses is equivalent to (7.6).

An interesting property is that the normal form of $\mathsf{SCP}(\mathsf{SRed}(cL \geq d))$ consists of the set of all prime extended clauses of $cL \geq d$.

Proposition 7.1.7 *For all extended clauses $L' \geq d'$ dominated by $cL \geq d$ there exists an extended clause in $\mathsf{SCP}(\mathsf{SRed}(cL \geq d))$ dominating $L' \geq d'$.*

Proof. Suppose that $L' \geq d'$ is dominated by $\mathsf{SCP}(\mathsf{SRed}(cL \geq d))$. Because $\mathsf{Ext}(cL \geq d) = \mathsf{Ext}(\mathsf{SCP}(\mathsf{SRed}(cL \geq d)))$, we know that $cL \geq d$ dominates $L' \geq d'$. Obviously, $cL \geq d$ dominates $L' \geq d'$ only if it dominates $L'' \geq d'$ with $L'' = L' \cap L$. We consider the strict reduction $c''L'' \geq d''$ of $cL \geq d$ and its corresponding strongest extended clause $L'' \geq \beta$. From Proposition 7.1.2 we know that $\beta \geq d'$, and therefore $L'' \geq \beta$ in $\mathsf{SCP}(SRed(cL \geq d))$ dominates $L' \geq d'$. □

For a single linear pseudo-Boolean inequality $cL \geq d$ the normal form of $\mathsf{SCP}(\mathsf{SRed}(cL \geq d))$ is already the *ideal solved form* that should be extracted by our constraint solver. Note that if there are two or more linear pseudo-Boolean inequalities, then the union of the sets of generated extended clause does not contain all prime extended clauses of the set of these linear pseudo-Boolean inequalities. We then have to apply *Generalized Resolution* in order to obtain the ideal solved form.

The challenge is to find the normal form of $\mathsf{SCP}(\mathsf{SRed}(cL \geq d))$ without completely constructing the dominated and therefore *redundant* extended clauses, and to minimize the number of domination checks. In the following we describe strong redundancy criteria identifying extended clauses that are dominated by other extended clauses.

7.2 IDENTIFYING REDUNDANT EXTENDED CLAUSES

The normal form of SCP(SRed($cL \geq d$)), i.e., the set of all non-redundant extended clauses in SCP(SRed($cL \geq d$)), is the set of extended clauses, equivalent to $cL \geq d$, that we want to build. A naive way is to completely generate SCP(SRed($cL \geq d$)), and then delete the redundant extended clauses. Typically, many of the generated extended clauses are redundant. We develop strong redundancy criteria that minimize the number of domination checks needed for deciding whether an extended clause is redundant, i.e., is dominated by another extended clause. We show how to decide redundancy of an extended clause in SCP(SRed($cL \geq d$)) by checking domination against *two* other extended clauses. For this purpose we first arrange the elements of SCP(SRed($cL \geq d$)) as a directed acyclic graph.

Let $T = T(cL \geq d)$ be a finite acyclic graph with root node, where the nodes are labeled with the strict reductions of $cL \geq d$. The root of T is labeled by $cL \geq d$. The direct subgraphs of T are the graphs $T(c'L' \geq d')$ for all strict reductions $c'L' \geq d'$ of $cL \geq d$, where exactly one literal has been eliminated. We denote by $\mathcal{T}(T, j)$ the direct subgraph of T with the label $\sum_{1 \leq i \neq j \leq n} c_i L_i \geq d - c_j$, provided that $d - c_j \geq 1$. We abbreviate $\mathcal{T}(\mathcal{T}(\ldots(\mathcal{T}(T, j_1), j_2), \ldots), j_k)$ by $T(\{j_1, \ldots, j_k\})$. We say $j_1 \ldots j_k$ is a *path* to the graph $T(\{j_1, \ldots, j_k\})$. Note that every permutation of $j_1 \ldots j_k$ is a path to $T(\{j_1, \ldots, j_k\})$, and that every element of SRed($cL \geq d$) is the label of exactly one node in $T(cL \geq d)$. We denote by T_j the strongest extended clause of the root label of $\mathcal{T}(T, j)$. The strongest extended clause of the root label of $T(\{j_1, \ldots, j_k\})$ is referenced by $T_{\{j_1, \ldots, j_k\}}$. Note that the extended clause $T_{\{j_1, \ldots, j_k\}}$ has $n - k$ literals, therefore the length of the path determines the number of literals of the label. A part of the graph view of SCP(SRed($cL \geq d$)) is shown in Figure 7.1. For each node we show the strict reduction and its corresponding strongest extended clause. Additionally, the arcs are labeled by the literal L_i that we eliminated from the father node to the son node.

In the following we denote by $J_k := \{j_1, \ldots, j_k\} \subset \{1, \ldots, n\}$ a set of indices such that $T(J_k)$ is a valid strict reduction of $cL \geq d$. Furthermore, we denote by $\overline{J_k} := \{j'_1, \ldots, j'_{n-k}\}$ the set of indices of the remaining literals in $T(J_k)$. Thus, $\overline{J_k} \cup J_k = \{1, \ldots, n\}$ and $\overline{J_k} \cap J_k = \emptyset$. We then have

$$T(J_k) = \sum_{i=1}^{n-k} c_{j'_i} L_{j'_i} \geq d - \sum_{i=1}^{k} c_{j_i} .$$

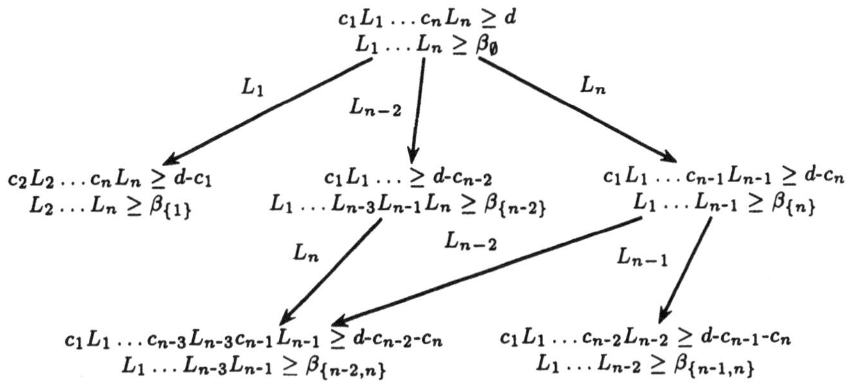

Figure 7.1 Part of a graph view of SCP(SRed($cL \geq d$))

We abbreviate $\sum_{i=1}^{n-k} c_{j'_i} L_{j'_i}$ by $c(J_k)L(J_k)$ and $\sum_{i=1}^{n-k} L_{j'_i}$ by $L(J_k)$. First, we investigate the degrees of the strongest extended clauses along a path.

Lemma 7.2.1 *For all* J_k *and* $J_{k+1} = J_k \cup \{j_{k+1}\}$ *we have*

$$\deg(T_{J_k}) \geq \deg(T_{J_{k+1}}) \geq \deg(T_{J_k}) - 1 \ . \tag{7.7}$$

Furthermore, for all $J_k \subset J_l$ *we have*

$$|J_l| - |J_k| \geq \deg(T_{J_k}) - \deg(T_{J_l}) \ . \tag{7.8}$$

Proof. First, we show (7.7). Let T_{J_k} be $L_{j_{k+1}} + L(J_{k+1}) \geq \beta$ and let $T_{J_{k+1}}$ be $L(J_{k+1}) \geq \beta'$. We need to show that $\beta \geq \beta' \geq \beta - 1$.

$\beta \geq \beta'$: We know that $L(J_{k+1}) \geq \beta'$ dominates $L_{j_{k+1}} + L(J_{k+1}) \geq \beta'$. Since $L_{j_{k+1}} + L(J_{k+1}) \geq \beta$ is the strongest extended clause, we have $\beta \geq \beta'$.

$\beta' \geq \beta - 1$: We know that $L_{j_{k+1}} + L(J_{k+1}) \geq \beta$ dominates its own reduction $L(J_{k+1}) \geq \beta - 1$. Because $L(J_{k+1}) \geq \beta'$ is a strongest extended clause, we have $\beta' \geq \beta - 1$.

Therefore, the degree of the strongest extended clauses along a path from the root at each step either decreases by 1 or stays equal. By induction, we conclude that (7.8) holds for all $J_k \subset J_l$. □

Lemma 7.2.2 *Let* $T(J_k)$ *be* $c_{j_{k+1}} L_{j_{k+1}} + \sum_{i=1}^{n-(k+1)} c_{j'_i} L_{j'_i} \geq d' + c_{j_{k+1}}$. *W.l.o.g. we assume that* $c_{j'_p} \geq c_{j'_q}$ *for all* $1 \leq p < q \leq n-(k+1)$. *Let* $\deg(T_{J_k}) = \beta$ *and* $\deg(T_{J_{k+1}}) = \beta'$. *If* $c_{j_{k+1}} \geq c_{j'_{\beta'}}$, *then* $\beta' + 1 = \beta$.

Proof. The strict reduction $T(J_{k+1})$ of $T(J_k)$ is $\sum_{i=1}^{n-(k+1)} c_{j'_i} L_{j'_i} \geq d'$. From equation (7.4) in Proposition 7.1.2 we know that

$$\sum_{i=1}^{\beta'-1} c_{j'_i} < d' \leq \sum_{i=1}^{\beta'} c_{j'_i} \ .$$

By adding $c_{j_{k+1}}$ to the left inequality, we derive

$$c_{j_{k+1}} + \sum_{i=1}^{\beta'-1} c_{j'_i} < d' + c_{j_{k+1}} \ . \tag{7.9}$$

According to Proposition 7.1.2,. β is the smallest number of the largest coefficients such that the sum of the largest coefficients is greater than the right-hand side $d' + c_{j_{k+1}}$. From $c_{j_{k+1}} \geq c_{j'_{\beta'}}$ and (7.9) we know that $c_{j_{k+1}}$ is one of the coefficients of this sum, and therefore $\beta > \beta'$. By Lemma 7.2.1 we have $\beta \leq \beta' + 1$, and therefore $\beta' + 1 = \beta$. □

The converse does not necessarily hold.

Example 7.2.3 Let the strict reduction $T(J_{k+1})$ be

$$6 \cdot L_1 + 5 \cdot L_2 + 4 \cdot L_3 + 3 \cdot L_4 \geq 12 \ .$$

Its strongest extended clause is

$$L_1 + L_2 + L_3 + L_4 \geq 3$$

with degree 3, because $6 + 5 < 12 \leq 6 + 5 + 4$. Let $c_i L_i$ be the product that has been eliminated, thus the inequality at the father node is

$$c_i \cdot L_i + 6 \cdot L_1 + 5 \cdot L_2 + 4 \cdot L_3 + 3 \cdot L_4 \geq 12 + c_i \ ,$$

and its strongest extended clause is

$$L_i + L_1 + L_2 + L_3 + L_4 \geq \beta \ .$$

By Lemma 7.2.2 we know that if $c_i \geq 4$ then $\beta = 4$. Suppose that $c_i = 4$, then condition (7.4) becomes $6 + 5 + 4 < 16 \leq 6 + 5 + 4 + 4$ and, as expected, the degree β is 4. Note that if $c_i = 3$, then β is still 4.

We now give a complete redundancy criterion for $\mathsf{SCP}(cL \geq d) = T_\emptyset$.

Lemma 7.2.4 *$T_\emptyset$ is redundant if and only if* $\deg(T_\emptyset) = \deg(T_{\{n\}})$.

Proof. We show that $T_\emptyset$ is dominated by a strongest extended clause if and only if it is dominated by $T_{\{n\}}$. Let $T_\emptyset$ be $L \geq \beta$, let L' be $L \setminus \{L_n\}$ and let $T_{\{n\}}$ be $L' \geq \beta'$. Note that $c_n \leq c_i$ for all $1 \leq i \leq n$.

$\Leftarrow$: Since $\beta = \beta'$, we know that $L' \geq \beta$ dominates $L \geq \beta$ because $L' \subset L$. Hence, $L \geq \beta$ is redundant.

$\Rightarrow$: Let us assume that $L \geq \beta$ is redundant. Then there exists $L'' \geq \beta''$ in $\mathsf{SCP}(\mathsf{SRed}(cL \geq d))$ that dominates $L \geq \beta$, and thus $\beta'' \geq \beta$. Since $L'' \geq \beta''$ is the strongest extended clause of a strict reduction of $cL \geq d$, we have by Lemma 7.2.1 that $\beta \geq \beta''$ and therefore $\beta = \beta''$. Because the degree of the strongest extended clauses along a path always decreases by 1 or stays equal, see (7.8), and since $L'' \subset L$, we know that all strongest extended clauses along the path from the root to the node with strongest extended clause $L'' \geq \beta$ have the degree β. Thus, for some direct son of the root we have a strongest extended clause $L \setminus \{L_j\} \geq \beta$ dominating $L \geq \beta$. From Lemma 7.2.2 we know that the coefficient c_j of the eliminated literal L_j is smaller than $c_{\beta-1}$, because otherwise $\beta'' + 1 = \beta + 1 = \beta$. From $c_j < c_{\beta-1}$ we derive $j \geq \beta - 1$, since $cL \geq d$ is in pseudo-Boolean normal form and therefore $\sum_{i=1}^{\beta-1} c_i < d - c_j$. Because $c_j \geq c_n$, we have $d - c_j \leq d - c_n$, and therefore $\sum_{i=1}^{\beta-1} c_i < d - c_n$. But then $\beta' \geq \beta$, and because β' is the degree of a strongest extended clause on a path from the root, we have $\beta' \leq \beta$ and therefore $\beta' = \beta$. □

Proposition 7.2.5 *If T_{J_k} is dominated by T_{J_l} and $J_l \supset J_k$, then $T_{J_{k+1}} := T_{J_k \cup \{j_{k+1}\}}$ dominates T_{J_k} with $c_{j_{k+1}} \leq c_i$ for all $i \in \overline{J_k}$.*

Proof. The proposition follows immediately from Lemma 7.2.4 if we view $T(J_k)$ as an independent graph. The specific son is then the one, where the literal with the smallest coefficient has been eliminated. □

With Proposition 7.2.5 we can decide whether T_{J_k} is dominated by an extended clause in a subgraph of the graph with root label $T(J_k)$ by checking domination against one specific extended clause, i.e., the strongest extended clause from the strict reduction of $T(J_k)$, where we have eliminated the literal with the smallest coefficient. In Figure 7.2 the basic idea of Lemma 7.2.4 is illustrated.

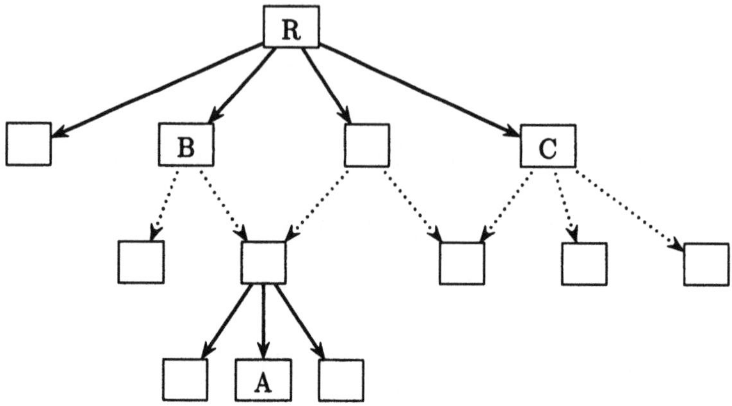

Figure 7.2 Redundancy caused by an extended clause in the subgraph

The graph corresponds to a part of the graph view of $\mathsf{SCP}(\mathsf{SRed}(cL \geq d))$. Assume that the sons of each node are ordered such that the coefficient of the eliminated literal of a son node is greater than or equal to all the coefficients of the eliminated literals of its right brothers. Now Lemma 7.2.4 says that there is an extended clause A under the extended clause R dominating R if and only if C dominates R. In the proof of Lemma 7.2.4 we show that if there is an A dominating R, then B is dominating R, from which we show that C is dominating R. The key idea is to show that the strongest extended clauses A,B, and C all have the same degree.

We now give a similar criterion for extended clauses lying on a path to $T(J_k)$, and, as in Proposition 7.2.5, identify a single specific extended clause that needs to be considered.

Lemma 7.2.6 *For $J_p \subset J_k$ the extended clause T_{J_k} is dominated by T_{J_p} if and only if*

$$|J_k| - |J_p| \leq \deg(T_{J_p}) - \deg(T_{J_k}) \ .$$

Proof. Because $|\{1,\ldots,n\} \setminus J_p| - |\{1,\ldots,n\} \setminus J_k| = |J_k| - |J_p|$, the lemma follows immediately from Lemma 3.1.14. □

Lemma 7.2.7 *If $T_{J_{k+1}}$ is dominated by T_{J_p} with $J_p \subset J_{k+1}$, then $T_{J_{k+1}\setminus\{j_i\}}$ dominates $T_{J_{k+1}}$ for all $j_i \in J_{k+1} \setminus J_p$.*

Proof. From (7.8) we know that $|J_{k+1}| - |J_p| \geq \deg(T_{J_p}) - \deg(T_{J_{k+1}})$ and from Lemma 7.2.6 we know that $|J_{k+1}| - |J_p| \leq \deg(T_{J_p}) - \deg(T_{J_{k+1}})$, therefore we have $|J_{k+1}| - |J_p| = \deg(T_{J_p}) - \deg(T_{J_{k+1}})$. Since in this case the degree of all extended clauses on a path from T_{J_p} to $T_{J_{k+1}}$ decreases always by one, we know that the degree of all direct fathers, $T_{J_{k+1}\setminus\{j_i\}}$ for all $j_i \in J_{k+1} \setminus J_p$, is the degree of $T_{J_{k+1}}$ plus one. Thus, for all $j_i \in J_{k+1} \setminus J_p$ we have $\deg(T_{J_{k+1}\setminus\{j_i\}}) - \deg(T_{J_{k+1}}) = 1$ and since $|J_{k+1}| - |J_{k+1} \setminus \{j_i\}| = 1$, we have domination by Lemma 7.2.6. □

We identify similar to Proposition 7.2.5 a single specific extended clause that needs to be tested for application of Lemma 7.2.6.

Proposition 7.2.8 *If $T_{J_{k+1}}$ is dominated by T_{J_p} with $J_p \subset J_{k+1}$, then there exists $T_{J_{k+1}\setminus\{j_{\max}\}}$ that dominates $T_{J_{k+1}}$, where $c_{j_{\max}} \geq c_{j_i}$ for all $1 \leq i \leq k+1$.*

Proof. Let us consider a specific path $H_{k+1} := h_1 \ldots h_{k+1}$ from the root to $T(J_{k+1})$, where $c_{h_1} \leq \cdots \leq c_{h_{k+1}} = c_{j_{\max}}$ and $h_{k+1} = j_{\max}$, i.e., we always eliminate the literal with the smallest coefficient first. For all direct fathers of $T(J_{k+1})$ the coefficient of the literal that is eliminated is therefore less than or equal to $c_{h_{k+1}}$. Let $\beta = \deg(T_{J_{k+1}})$. Because of Lemma 7.2.7, we have a strict reduction

$$R := \sum_{i \in \overline{J_{k+1}}} c_i L_i + c_x L_x \geq d + c_x - \sum_{i \in J_{k+1}} c_i$$

with $x \in J_p$ and $\deg(\mathsf{SCP}(R)) = \beta + 1$. Using our fixed path H_{k+1}, we know that there is another strict reduction

$$H := \sum_{i \in \overline{J_{k+1}}} c_i L_i + c_{j_{\max}} L_{j_{\max}} \geq d + c_{j_{\max}} - \sum_{i \in J_{k+1}} c_i \ .$$

We now show that $\deg(\mathsf{SCP}(R)) = \deg(\mathsf{SCP}(H)) = \beta + 1$.

Suppose that $c_{j_{\max}} \geq c_{j'_\beta}$. As in Lemma 7.2.2 we derive that $\deg(\mathsf{SCP}(H)) > \beta$, hence $\deg(\mathsf{SCP}(H)) = \beta + 1$.

Now suppose that $c_{j_{\max}} < c_{j'_\beta}$. Since we have $c_{j_{\max}} \geq c_x$, we also have $c_x < c_{j'_\beta}$. Because $\deg(\mathsf{SCP}(R)) = \beta + 1$, we have

$$\sum_{1 \leq i \leq \beta} c_{j'_i} < d + c_x - \sum_{i \in J_{k+1}} c_i \ .$$

With $c_{j_{\max}} \geq c_x$ we get

$$\sum_{1 \leq i \leq \beta} c_{j'_i} < d + c_{j_{\max}} - \sum_{i \in J_{k+1}} c_i \ ,$$

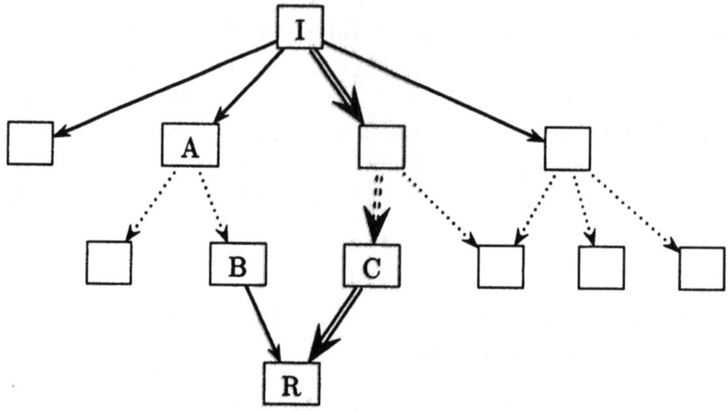

Figure 7.3 Redundancy caused by an extended clause above in the graph

which also implies deg(SCP(H)) $> \beta$ and therefore deg(SCP(H)) $= \beta + 1$. Let $J_{k+1} \setminus \{j_{\max}\}$ be $\{h_1, \ldots, h_k\}$, then the proposition follows. □

We illustrate the basic idea of Proposition 7.2.8 in Figure 7.3. Proposition 7.2.8 says that there is an A dominating R if and only if C dominates R. In the graph we assume that the fathers of each node are ordered such that the coefficient of the eliminated literal to come from a father node to a son node is greater than or equal to all the coefficients of the eliminated literals of its *left* neighbors. In the proof of Proposition 7.2.8 we show that if A dominates R, then there is a B dominating R, and that all degrees along a path from R to A decrease by one. The double line corresponds to the specific path H_{k+1}, and we show that C, lying on this path, has the same degree as B and therefore dominates R.

We combine now Proposition 7.2.5 and Proposition 7.2.8 to a complete redundancy criterion.

Theorem 7.2.9 *The strongest extended clause $T_{J_{k+1}}$ is redundant if and only if*

[a] $T_{J_{k+1} \cup \{j_{k+2}\}}$ *dominates* $T_{J_{k+1}}$ *with* $c_{j_{k+2}} \leq c_i$ *for all* $i \in \overline{J_{k+1}}$, *or*

[b] $T_{J_{k+1} \setminus \{j_{\max}\}}$ *dominates* $T_{J_{k+1}}$ *with* $c_{j_{\max}} \geq c_i$ *for all* $i \in J_{k+1}$.

Proof.

$\Leftarrow$: [a] from Proposition 7.2.5 and [b] from Proposition 7.2.8.

$\Rightarrow$: Suppose that neither [a] nor [b] is satisfied for $T_{J_{k+1}}$, but $T_{J_{k+1}}$ is redundant. Then there is a strongest extended clause $L'' \geq \beta$ that dominates $T_{J_{k+1}} := L' \geq \beta'$ and $L'' \neq L'$. From Proposition 7.2.8 we have $L'' \not\supset L'$ and from Proposition 7.2.5 we have $L'' \not\subset L'$. Therefore, there is at least one literal in L'' that is not in L'. Following the proof of Lemma 3.1.14 a valid reduction of $L'' \geq \beta$ is $L'' \cap L' \geq \beta - |L'' \setminus L'|$, which also dominates $L' \geq \beta'$. The strongest extended clause containing exactly the literals in $L'' \cap L'$ therefore dominates $L \geq \beta'$ and hence $T_{J_{k+1}}$. Since $L'' \cap L' \subset L''$, because $L'' \setminus L' \neq \emptyset$, Proposition 7.2.5 applies, and therefore condition [a]. □

For illustrating statement [b] of Theorem 7.2.9 we look again at Figure 7.2. We have shown that if there is an X dominating R and X is not on a path also including R, then we will find an A in the subgraph under X such that Proposition 7.2.5 applies.

By Theorem 7.2.9 we can decide redundancy of a strongest extended clause by investigating exactly 2 other strongest extended clauses. We end this section with an example.

Example 7.2.10 Let us transform the linear pseudo-Boolean inequality

$$5 \cdot L_1 + 4 \cdot L_2 + 3 \cdot L_3 + 2 \cdot L_4 + 1 \cdot L_5 \geq 11 \ .$$

Its strongest extended clause is

$$L_1 + L_2 + L_3 + L_4 + L_5 \geq 3 \ .$$

Let us compute the strict reductions $T(\{i\})$ and their corresponding strongest extended clauses $T_{\{i\}}$.

i	$T(\{i\})$	$T_{\{i\}}$
1	$4 \cdot L_2 + 3 \cdot L_3 + 2 \cdot L_4 + 1 \cdot L_5 \geq 6$	$L_2 + L_3 + L_4 + L_5 \geq 2$
2	$5 \cdot L_1 + 3 \cdot L_3 + 2 \cdot L_4 + 1 \cdot L_5 \geq 7$	$L_1 + L_3 + L_4 + L_5 \geq 2$
3	$5 \cdot L_1 + 4 \cdot L_2 + 2 \cdot L_4 + 1 \cdot L_5 \geq 8$	$L_1 + L_2 + L_4 + L_5 \geq 2$
4	$5 \cdot L_1 + 4 \cdot L_2 + 3 \cdot L_3 + 1 \cdot L_5 \geq 9$	$L_1 + L_2 + L_3 + L_5 \geq 2$
5	$5 \cdot L_1 + 4 \cdot L_2 + 3 \cdot L_3 + 2 \cdot L_4 \geq 10$	$L_1 + L_2 + L_3 + L_4 \geq 3$

We see that $T_{\{5\}}$ dominates $T_\emptyset$ (Theorem 7.2.9[a] $\Leftarrow$) and that $T_{\{1\}}, T_{\{2\}}, T_{\{3\}}$, and $T_{\{4\}}$ are dominated by $T_\emptyset$ (Theorem 7.2.9[b] $\Leftarrow$). We calculate $T_{\{4,5\}} =$

$L_1+L_2+L_3 \geq 2$ and conclude that $T_{\{5\}}$ is not redundant(Theorem 7.2.9; completeness). Proceeding further, we arrive at the set of non-redundant extended clauses given in Example 7.1.6.

7.3 IMPLEMENTATION

We sketch an implementation of the transformation algorithm using the strong redundancy criteria presented in Section 7.2.

Theorem 7.2.9 is a complete redundancy criterion for a strongest extended clause that reduces the number of needed domination checks to 2. We denote by $\beta' = \deg(T_{J_k \setminus \{j_{max}\}})$ the degree of the strongest extended clause of the specific father of $T(J_{k+1})$ and by $\beta'' = \deg(T_{J_{k+1} \cup \{j_{k+2}\}})$ the degree of the strongest extended clause of the specific son of $T(J_{k+1})$. The degree of the extended clause for which we want to check redundancy is denoted by $\beta = \deg(T_{J_{k+1}})$. From Theorem 7.2.9 we see immediately that $T_{J_{k+1}}$ is redundant if and only if

$$\beta' - 1 = \beta \qquad \text{or} \qquad \beta = \beta'' \ .$$

Hence, for deciding redundancy of a strongest extended clause only the degrees of two strongest extended clauses (the specific father and the specific son) are needed. We describe an algorithm that computes from a linear pseudo-Boolean inequality $cL \geq d$ in normal form a set of equivalent non-redundant extended clauses. We assume that $\sum_{i=1}^{n} c_i \geq d$, since otherwise $cL \geq d$ is unsatisfiable. We also assume that $\sum_{i=2}^{n} c_i \geq d$, since otherwise $\alpha(L_1) = 1$ for all $\alpha \in \mathsf{Ext}(cL \geq d)$, and we can consider the simpler problem of transforming

$$\sum_{i=2}^{n} c_i L_i \geq d - c_1 \ .$$

In the following we assume that the linear pseudo-Boolean inequality $cL \geq d$ is satisfiable, in pseudo-Boolean normal form and that for every literal $L_i \in L$ there are assignments $\alpha_1, \alpha_2 \in \mathsf{Ext}(cL \geq d)$ such that $\alpha_1(L_i) = 1 - \alpha_2(L_i)$. Note that for all generated strongest extended clauses $L' \geq \beta$, we have $|L'| > \beta$, since otherwise $\alpha(L_i) = 1$ for all $L_i \in L'$, which we have excluded. We can always ensure this condition by applying pseudo-Boolean unit resolution in advance.

In the program fragments we use the following non-standard notation. Given a linear pseudo-Boolean term cL, we select the last (resp. first) coefficient/literal pair $c_l L_l$ and the remaining term $c'L'$ by $cL = c'L' + c_l L_l$ (resp. $cL = c_l L_l + c'L'$).

First, we give the procedure **get_beta** calculating the degree β of the strongest extended clause of its argument, a strict reduction, and β'', the degree of the strongest extended clause of the strict reduction, where we have eliminated the literal with the smallest coefficient. The right-hand side β'' is then the degree of the strongest extended clause of the specific son node as in Proposition 7.2.8.

Algorithm 7.3.1 *Get greatest possible right-hand side*

```
get_beta(cL ≥ d)
        (i, β, sum, prevsum) := (1, 0, 0, 0)
        while sum < d
                cL = c_l L_l + c'L'
                prevsum := sum
                (sum, β, cL) := (sum + c_l, β + 1, c'L')
        endwhile
        while cL = c_l L_l + c'L'
                cL := c'L'
        endwhile
        (* c_l is the smallest coefficient *)
        β'' := if prevsum ≥ d − c_l then β − 1 else β endif
        return (β, β'')
end get_beta
```

We obtain β by summing up the coefficients according to (7.3), and count the number of added coefficients until we reach the right-hand side. Since the specific son is the one where the literal with the smallest coefficient has been eliminated, this coefficient cannot be part of the sum. Note that $\sum_{i=2}^{n} c_i \geq d$. Therefore, the degree of the specific son is one smaller if and only if the current sum minus the last added coefficient, i.e., the previous sum, is greater than or equal to the right-hand side $d - c_l$ of the specific son. Hence, the overhead for calculating β'' is not high. In the actual implementation we store the value of the smallest coefficient for each strict reduction and so avoid searching for it.

In an efficient implementation we do not generate the whole graph. Instead, we implicitly visit the nodes, i.e., the strict reductions, in a specific order such that for each strict reduction the redundancy test becomes trivial. We ensure that each strict reduction $T(J_{k+1})$ is visited only once through the path H_{k+1} (see Proof of Proposition 7.2.8). This is achieved by splitting the linear pseudo-Boolean term (the left-hand side of the linear pseudo-Boolean inequality) into two parts, one containing the literals that must be eliminated and the other one containing the literals that need not be eliminated. We eliminate literals having a smaller coefficient first and then forbid further reduction on these literals by moving them from the first part to the second part. Thus, we make sure that the strict reductions are visited only through the specific path. We give a recursive procedure `transform` generating each possible strict reduction only once and adding its strongest extended clause to the output set if and only if it is not redundant. As parameters we have a linear pseudo-Boolean inequality, which is split into two parts and the right-hand side, and the right-hand side β' of the strongest extended clause of the father in the path H_{k+1}.

Algorithm 7.3.2 *Transform a linear pseudo-Boolean inequality*

`transform`$(cL, \hat{c}\hat{L}, d, \beta')$
  $S := \emptyset$
  $(\beta, \beta'') =$ **get_beta**$(cL + \hat{c}\hat{L} \geq d)$
  if $\beta' - 1 \neq \beta \wedge \beta'' \neq \beta$ **then** (* not redundant *)
    $S := S \cup \{L + \hat{L} \geq \beta\}$ (* add it *)
  endif
  (* cL is non-empty and there are valid strict reductions *)
  while $c'L' + c_lL_l = cL \wedge d - c_l \geq 1$ (* eliminate c_lL_l *)
    $S := S \cup$ `transform`$(c'L', \hat{c}\hat{L}, d - c_l, \beta)$
    $\hat{c}\hat{L} := \hat{c}\hat{L} + c_lL_l$ (* forbid further reduction on L_l *)
    $cL := c'L'$
  endwhile
  return S
end `transform`

Note that in the **while**-loop we select first the product with the smallest coefficient and then forbid further reduction on it by moving the product to the second part. The set of all non-redundant strongest extended clauses equivalent to the linear pseudo-Boolean inequality $cL \geq d$ is `transform`$(cL, \emptyset, d, 0)$. We finish this section with some examples.

Example 7.3.3 The examples are taken from [NW88].

[a] [NW88, page 266] Let $cL \geq d$ be

$$79 \cdot X_1 + 53 \cdot X_2 + 53 \cdot X_3 + 45 \cdot X_4 + 45 \cdot X_5 \leq 178 \ .$$

The pseudo-Boolean normal form of $cL \geq d$ is

$$79 \cdot \overline{X_1} + 53 \cdot \overline{X_2} + 53 \cdot \overline{X_3} + 45 \cdot \overline{X_4} + 45 \cdot \overline{X_5} \geq 97 \ .$$

The equivalent set of extended clauses is

$$\begin{array}{rcl} \{\overline{X_1} + \overline{X_2} + \overline{X_3} + \overline{X_4} + \overline{X_5} & \geq & 2, \\ \overline{X_1} + \overline{X_2} + \overline{X_3} & \geq & 1\} \ . \end{array}$$

[b] [NW88, page 460] Let $cL \geq d$ be

$$\begin{array}{l} 774 \cdot X_1 + 76 \cdot X_2 + 22 \cdot X_3 + 42 \cdot X_4 + 21 \cdot X_5 + \\ 760 \cdot X_6 + 818 \cdot X_7 + 62 \cdot X_8 + 785 \cdot X_9 \leq 1500 \ . \end{array}$$

The transformation procedure generates the single extended clause

$$\overline{X_1} + \overline{X_6} + \overline{X_7} + \overline{X_9} \geq 3 \ ,$$

which is equivalent to $cL \geq d$.

[c] [NW88, page465] Let $cL \geq d$ be

$$\begin{array}{l} 300 \cdot X_3 + 300 \cdot X_4 + 285 \cdot X_5 + 285 \cdot X_6 + 265 \cdot X_8 + 265 \cdot X_9 + \\ 230 \cdot X_{12} + 230 \cdot X_{13} + 190 \cdot X_{14} + 200 \cdot X_{22} + 400 \cdot X_{23} + \\ 200 \cdot X_{24} + 400 \cdot X_{25} + 200 \cdot X_{26} + 400 \cdot X_{27} + 200 \cdot X_{28} + \\ 400 \cdot X_{29} + 200 \cdot X_{30} + 400 \cdot X_{31} \leq 2700, \end{array}$$

which is a linear 0-1 inequality from the constraint set of the 0-1 integer programming problem p0033 [CJP83, BBI92]. Transforming $cL \geq d$ generates 4282 non-redundant strongest extended clauses. They have been calculated with a PROLOG-implementation of the algorithm in 4.8 seconds cpu-time on a SPARC-10/31. All 15 inequalities of the problem produce together 8710 extended clauses, where 749 extended clauses are dominated by extended clauses generated from another inequality such that 7961 extended clauses remain. Note that the number of non-redundant classical clauses that are equivalent to $cL \geq d$ is 117520.

Example 7.3.3[c] demonstrates the advantage of using extended clauses instead of classical clauses. Because of the large number of classical clauses needed to represent a linear pseudo-Boolean inequality, clausal satisfiability methods cannot be applied to typical 0-1 integer programming problems [GG80]. The reformulation of a linear pseudo-Boolean inequality as a set of extended clauses is more likely to be applicable to general 0-1 problems because of its more compact representation.

In the next section we give an even more compact representation of linear pseudo-Boolean inequalities and extended clauses, which allows us to speed up the computation and brings even larger problems into the scope of this symbolic method.

7.4 SYMMETRIES

We introduce the concept of symmetries for a set of literals w.r.t. a linear pseudo-Boolean inequality and then define the *compact set* representation of linear pseudo-Boolean inequalities and extended clauses. This more compact formulation helps to avoid *redundant computations* in the transformation method.

Let us transform the linear pseudo-Boolean inequality

$$5 \cdot A + 5 \cdot B + 4 \cdot C + 3 \cdot D + 3 \cdot E + 3 \cdot F \geq 12 \ . \tag{7.10}$$

For example, three strict reductions of (7.10) are

$$5 \cdot A + 5 \cdot B + 4 \cdot C + 3 \cdot E + 3 \cdot F \geq 9 \tag{7.11}$$

$$5 \cdot A + 5 \cdot B + 4 \cdot C + 3 \cdot D + 3 \cdot F \geq 9 \tag{7.12}$$

$$5 \cdot A + 5 \cdot B + 4 \cdot C + 3 \cdot D + 3 \cdot E \geq 9 \tag{7.13}$$

obtained by eliminating D for (7.11), E for (7.12) and F for (7.13). Note that these three strict reductions are identical, except for the names of the last two literals. Therefore, these three extended clauses yield exactly the same set of non-redundant extended clauses except for the names of the literals. We say that D, E and F are *symmetric* in (7.10). The study of symmetries for propositional calculus problems [BS92] is not directly related to the presented approach. Although the main idea, i.e., avoiding redundant computation by exploiting symmetry properties, is similar. In the following we describe how to avoid redundant computations occurring while eliminating symmetric literals. For this purpose we first define a compact representation of a set of linear pseudo-Boolean inequalities.

Definition 7.4.1 For all $1 \leq i < j \leq m$ let S_i and S_j be sets of literals such that $\mathsf{Var}(L_l) \neq \mathsf{Var}(L_k)$ for all $L_l \in S_i$ and $L_k \in S_j$. We say that

$$\sum_{i=1}^{m} c_i S_i^{k_i} \geq d \ , \tag{7.14}$$

where $d \geq c_1 > \cdots > c_m \geq 1$ and $1 \leq k_i \leq |S_i|$ for all $1 \leq i \leq m$, is a *compact pseudo-Boolean inequality set* of a set of linear pseudo-Boolean inequalities. The compact pseudo-Boolean inequality set (7.14) represents the set of all linear pseudo-Boolean inequalities

$$\mathsf{expand}(\sum_{i=1}^{m} c_i S_i^{k_i} \geq d) := \left\{ cL \geq d \;\middle|\; \begin{array}{l} |L| = \sum_{i=1}^{m} k_i, \\ \forall c_l L_l \in cL : \exists S_i : L_l \in S_i \wedge c_l = c_i, \\ |S_i \cap L| = k_i \text{ for all } 1 \leq i \leq m \end{array} \right\} . \tag{7.15}$$

In other words, all linear pseudo-Boolean inequalities, where from each set of literals S_i exactly k_i occur in the linear pseudo-Boolean inequality with coefficient c_i. We say we *expand* a compact pseudo-Boolean inequality set if we replace $\sum_{i=1}^{m} c_i S_i^{k_i} \geq d$ by $\mathsf{expand}(\sum_{i=1}^{m} c_i S_i^{k_i} \geq d)$. For a set S of compact pseudo-Boolean inequality sets we define $\mathsf{expand}(S) := \bigcup_{I \in S} \mathsf{expand}(I)$.

Example 7.4.2 We can represent the single linear pseudo-Boolean inequality (7.10) by the compact pseudo-Boolean inequality set

$$5 \cdot \{A, B\}^2 + 4 \cdot \{C\}^1 + 3 \cdot \{D, E, F\}^3 \geq 12 \ .$$

The three strict reductions (7.11), (7.12) and (7.13) can be represented by

$$5 \cdot \{A, B\}^2 + 4 \cdot \{C\}^1 + 3 \cdot \{D, E, F\}^2 \geq 9 \ .$$

Note that every single linear pseudo-Boolean inequality can be represented as compact pseudo-Boolean inequality set. Then for all sets of literals S_i in its compact representation, we have $k_i = |S_i|$.

The compact pseudo-Boolean inequality set representation is the key to exploit the symmetry property of literals in a linear pseudo-Boolean inequality. The possible strict reductions that lead to an identical search can now be represented by a single compact pseudo-Boolean inequality set. We define similar to Definition 7.4.1 a compact set representation for extended clauses.

Definition 7.4.3 For all $1 \leq i < j \leq m$ let S_i and S_j be sets of literals such that $\mathsf{Var}(L_l) \neq \mathsf{Var}(L_k)$ for all $L_l \in S_i$ and $L_k \in S_j$. We say that

$$\sum_{i=1}^{m} S_i^{k_i} \geq \beta \tag{7.16}$$

with $1 \leq k_i \leq |S_i|$ for all $1 \leq i \leq m$, is a *compact extended clause set* of a set of extended clauses. The compact extended clause set (7.16) represents the set of all extended clauses

$$\mathsf{expand}(\sum_{i=1}^{m} S_i^{k_i} \geq \beta) := \left\{ L \geq \beta \;\middle|\; \begin{array}{l} |L| = \sum_{i=1}^{m} k_i \text{ and} \\ |S_i \cap L| = k_i \text{ for all } 1 \leq i \leq m \end{array} \right\} . \quad (7.17)$$

We say we *expand* a compact extended clause set $\sum_{i=1}^{m} S_i^{k_i} \geq \beta$, if we replace it by $\mathsf{expand}(\sum_{i=1}^{m} S_i^{k_i} \geq \beta)$. For a set S of compact extended clause sets we define $\mathsf{expand}(S) := \bigcup_{I \in S} \mathsf{expand}(I)$.

We generalize now the generation of strongest extended clauses of Proposition 7.1.2. For a compact pseudo-Boolean inequality set $\sum_{i=1}^{m} c_i S_i^{k_i} \geq d$ the corresponding strongest compact extended clause set is

$$\sum_{i=1}^{m} S_i^{k_i} \geq \beta \quad (7.18)$$

with

$$\sum_{i=1}^{\beta-1} c_i \cdot k_i < d \leq \sum_{i=1}^{\beta} c_i \cdot k_i \;. \quad (7.19)$$

Note that (7.19) corresponds to (7.3), but takes into account the compact set representation. We denote by $\mathsf{SCP}'(\sum_{i=1}^{m} c_i S_i^{k_i} \geq d) := (\sum_{i=1}^{m} S_i^{k_i} \geq \beta)$ the strongest compact extended clause set of a compact pseudo-Boolean inequality set. Obviously, for each linear pseudo-Boolean pseudo-Boolean inequality in $\mathsf{expand}(\sum_{i=1}^{m} c_i S_i^{k_i} \geq d)$ there is a corresponding strongest extended clause in $\mathsf{expand}(\sum_{i=1}^{m} S_i^{k_i} \geq \beta)$ and we have

$$\mathsf{expand}(\mathsf{SCP}'(\sum_{i=1}^{m} c_i S_i^{k_i} \geq d)) = \mathsf{SCP}(\mathsf{expand}(\sum_{i=1}^{m} c_i S_i^{k_i} \geq d)) \;.$$

A compact strict reduction of a compact pseudo-Boolean inequality set of the form $\sum_{i=1}^{m} c_i S_i^{k_i} \geq d$ is

$$\sum_{i=1}^{m} c_i S_i^{k_i'} \geq d - \sum_{i=1}^{m} c_i \cdot (k_i - k_i')$$

with either $k_i' = k_i$ or $k_i' = k_i - 1$ for all $1 \leq i \leq m$. Note that $c_i S_i^{k_i'}$ disappears if $k_i' = 0$. We denote by $\mathsf{SRed}'(\sum_{i=1}^{m} c_i S_i^{k_i} \geq d)$ the set of all compact strict reductions of a compact pseudo-Boolean inequality set. Obviously, we have

$$\mathsf{expand}(\mathsf{SRed}'(\sum_{i=1}^{m} c_i S_i^{k_i} \geq d)) = \mathsf{SRed}(\mathsf{expand}(\sum_{i=1}^{m} c_i S_i^{k_i} \geq d)) \ .$$

The transformation method using compact sets is now straightforward. For a compact pseudo-Boolean inequality set $\sum_{i=1}^{m} c_i S_i^{k_i} \geq d$ we have

$$\mathsf{expand}(\mathsf{SCP}'(\mathsf{SRed}'(\sum_{i=1}^{m} c_i S_i^{k_i} \geq d))) = \mathsf{SCP}(\mathsf{SRed}(\mathsf{expand}(\sum_{i=1}^{m} c_i S_i^{k_i} \geq d))) \ .$$

The practical advantage is that we avoid redundant computation of identical strict reductions modulo literal renaming because they are grouped together in a single compact pseudo-Boolean inequality set. Generalizing Theorem 7.2.9 for compact sets is immediate and an implementation of the transformation method using compact sets is similar to the implementation sketched in Section 5.4.

Example 7.4.4

[a] When transforming the linear pseudo-Boolean inequality

$$5 \cdot A + 5 \cdot B + 4 \cdot C + 3 \cdot D + 3 \cdot E + 3 \cdot F \geq 12 \ ,$$

we have to consider 32 strict reductions. If we transform the equivalent compact pseudo-Boolean inequality set

$$5 \cdot \{A, B\}^2 + 4 \cdot \{C\}^1 + 3 \cdot \{D, E, F\}^3 \geq 12 \ ,$$

then only 11 compact strict reductions are possible and the following set of compact extended clause sets is generated.

$$\begin{array}{rcl} \{\{A\}^1 + \{B\}^1 + \{C\}^1 + \{D\}^1 + \{E\}^1 + \{F\}^1 & \geq & 3, \\ \{A\}^1 + \{B\}^1 + \{C\}^1 + \{D, E, F\}^1 & \geq & 2\} \ . \end{array}$$

The set of equivalent strongest extended clauses is

$$\begin{array}{rcl} \{A + B + C + D + E + F & \geq & 3, \\ A + B + C + D & \geq & 2, \\ A + B + C + E & \geq & 2, \\ A + B + C + F & \geq & 2\} \ . \end{array}$$

So by exploiting the symmetry property of literals we avoid redundant computation and obtain a more compact result.

[b] The linear pseudo-Boolean inequality of Example 7.3.3[c] generates 4282 extended clauses in 4.8 seconds cpu time on a SPARC-10/31. The implementation using symmetries takes 0.77 seconds cpu time for generating the 4282 extended clauses, where 0.49 seconds are used to expand the 253 compact non-redundant extended clause sets. The transformation itself takes only 0.18 seconds.

The very compact representation of Example 7.4.4[b] suggests to further investigate the compact extended clause set representation of extended clauses. Generalizing symbolic methods for solving linear pseudo-Boolean problems, e.g., pseudo-Boolean unit resolution and *Generalized Resolution*, to work directly on the compact representation, brings even larger 0-1 problems in the scope of these methods. This issue will be investigated in the future.

7.5 DETECTION OF FIXED LITERALS

We present a method detecting some of the literals L_i for which the extended clause $L_i \geq 1$ is dominated by a set of linear pseudo-Boolean inequalities S. Given such an L_i, we have $\alpha(L_i) = 1$ for all assignments $\alpha \in \mathsf{Ext}(S)$. Hence, the literal L_i is fixed, and S can be simplified to a set of linear pseudo-Boolean inequalities S' not containing L_i or $\overline{L_i}$ such that

$$\mathsf{Ext}(S) = \mathsf{Ext}(S' \cup \{L_i \geq 1\}) \ .$$

A similar simplification is possible if several literals can be fixed. Let L be a set of literals such that S dominates $L_i \geq 1$ for all $L_i \in L$. We then can simplify S to a set of linear pseudo-Boolean inequalities S' not containing L_i or $\overline{L_i}$ for all $L_i \in L$ such that

$$\mathsf{Ext}(S) = \mathsf{Ext}(S' \cup \{L \geq |L|\}) \ .$$

Note that $L \geq |L|$ dominates $L_i \geq 1$ for all $L_i \in L$. We show how to find some of the fixed literals in L with help of pseudo-Boolean unit resolution of Chapter 5.

A subset of these literals that can be fixed is the set of unit literals $\mathtt{pbul}(S)$, defined on page 86. Pseudo-Boolean unit resolution is a procedure detecting this set of literals.

Example 7.5.1 Let S be

$$\begin{aligned}\{5 \cdot A + 4 \cdot B + 3 \cdot C + 2 \cdot D &\geq 10, \\ 8 \cdot \overline{A} + 6 \cdot D + 4 \cdot E + 3 \cdot F &\geq 8\}\end{aligned}$$

and let us follow a computation of $\texttt{pbur}(S)$, defined on page 86. First, we obtain $\texttt{fixed}(5 \cdot A + 4 \cdot B + 3 \cdot C + 2 \cdot D \geq 10) = A$ and simplify S to $\texttt{fix}(\texttt{A}, \texttt{S})$.

$$\begin{aligned}\{4 \cdot B + 3 \cdot C + 2 \cdot D &\geq 5, \\ 6 \cdot D + 4 \cdot E + 3 \cdot F &\geq 8\}\end{aligned}$$

Next, we have $\texttt{fixed}(6 \cdot D + 4 \cdot E + 3 \cdot F \geq 8) = D$ and obtain $\texttt{fix}(D, \texttt{fix}(A, S))$.

$$\begin{aligned}\{3 \cdot B + 3 \cdot C &\geq 3, \\ 2 \cdot E + 2 \cdot F &\geq 2\}\end{aligned}$$

Since now for all linear pseudo-Boolean inequalities $\texttt{fixed}$ returns $\perp$, we have

$$\mathsf{Ext}(S) = \mathsf{Ext}(\{3 \cdot B + 3 \cdot C \geq 3, 2 \cdot E + 2 \cdot F \geq 2, A + D \geq 2\}) \ .$$

Transforming the remaining linear pseudo-Boolean inequalities is trivial and we obtain

$$\mathsf{Ext}(S) = \mathsf{Ext}(\{B + C \geq 1, E + F \geq 1, A + D \geq 2\}) \ .$$

When a linear pseudo-Boolean inequality has been simplified by fixing literals, then transforming it is simpler and typically fewer strongest extended clauses are generated.

We know that a set of linear pseudo-Boolean inequalities S dominates $L_i \geq 1$ for some literal L_i if and only if $(S \cup \{\overline{L_i} \geq 1\})$ is unsatisfiable. We approximate the unsatisfiability test by checking whether the pseudo-Boolean unit relaxation of $S \cup \{\overline{L_i} \geq 1\}$ is unsatisfiable, i.e., whether $\perp \in \texttt{pbur}(S \cup \{\overline{L_i} \geq 1\})$. We get the procedure $\texttt{fixing}$ that detects some of the literals L_i for which $\alpha(L_i) = 1$ for all solutions α of a set of linear pseudo-Boolean inequalities S.

Algorithm 7.5.2 *Detect Fixed Literals*

```
fixing(S)
        K := ∅
        while ∃L_i : ⊥ ∈ pbur(S ∪ {¬L_i ≥ 1})
                K := K ∪ pbul(S ∪ {L_i ≥ 1})
                S := pbur(S ∪ {L_i ≥ 1})
        endwhile
        return K
end fixing
```

Note that S dominates $\texttt{fixing}(S) \geq |\texttt{fixing}(S)|$. Let K be $\texttt{fixing}(S)$, then

$$\mathsf{Ext}(S) = \mathsf{Ext}(\texttt{pbur}(S \cup \{K \geq |K|\}) \cup \{K \geq |K|\}) \ .$$

Note that **fixing** is a stronger procedure for detecting fixed literals of a set of linear pseudo-Boolean inequalities than pseudo-Boolean unit resolution and we have

$$\texttt{fixing}(S) \supseteq \texttt{pbul}(S) \ .$$

Example 7.5.3 Let S be

$$\begin{array}{rcl} \{5 \cdot A + 4 \cdot B + 3 \cdot C + 2 \cdot D & \geq & 7, \\ 3 \cdot \overline{B} + 3 \cdot D + 2 \cdot \overline{C} & \geq & 4\} \ , \end{array}$$

then none of the two linear pseudo-Boolean inequalities dominates $L_i \geq 1$ for some literal L_i. Let us apply **fixing** to S and start with calculating $\texttt{pbur}(S \cup \{\overline{A} \geq 1\})$. First, we obtain $\texttt{fix}(\overline{A}, S)$.

$$\begin{array}{rcl} \{4 \cdot B + 3 \cdot C + 2 \cdot D & \geq & 7, \\ 3 \cdot \overline{B} + 3 \cdot D + 2 \cdot \overline{C} & \geq & 4\} \end{array}$$

Now $\texttt{fixed}(4 \cdot B + 3 \cdot C + 2 \cdot D \geq 7) = B$ and we calculate $\texttt{fix}(B, \texttt{fix}(\overline{A}, S))$.

$$\begin{array}{rcl} \{3 \cdot C + 2 \cdot D & \geq & 3, \\ 3 \cdot D + 2 \cdot \overline{C} & \geq & 4\} \end{array}$$

Next, we derive $\texttt{fixed}(3 \cdot C + 2 \cdot D \geq 3) = C$ and $\texttt{fix}(C, \texttt{fix}(B, \texttt{fix}(\overline{A}, S)))$ gives

$$\begin{array}{rcl} \{ & & \top, \\ 3 \cdot D & \geq & 4\} \ , \end{array}$$

where $3 \cdot D \geq 4$ is $\perp$. We conclude that $S \cup \{\overline{A} \geq 1\}$ is unsatisfiable. Thus, S dominates $A \geq 1$ and we can replace S by $\texttt{fix}(A, S) \cup \{A \geq 1\}$.

$$\begin{array}{rcl} \{2 \cdot B + 2 \cdot C + 2 \cdot D & \geq & 2, \\ 3 \cdot \overline{B} + 3 \cdot D + 2 \cdot \overline{C} & \geq & 4, \\ A & \geq & 1\} \ . \end{array}$$

Note that $2 \cdot B + 2 \cdot C + 2 \cdot D \geq 2$ is already normalized. There is no further literal L_i such that $\perp \in \texttt{pbur}(\texttt{fix}(A, S) \cup \{A \geq 1\} \cup \{\overline{L_i} \geq 1\})$ and so **fixing** ends.

For example, on the constraint set of a 0-1 integer optimization problem "air01" found on MIPLIB [BBI92] **fixing** detects that 130 out of the 771 Boolean variables have to be fixed. But, `fixing` is a rather costly process. Suppose that n variables occur in S, then for a complete application of `fixing` we have to compute $2n$ pseudo-Boolean unit resolutions if `fixing` detects no fixed literals.

7.6 CONSTRAINED SIMPLIFICATION

In the context of constraint logic programming a linear pseudo-Boolean inequality $cL \geq d$ is typically part of a set of linear pseudo-Boolean inequalities S which are already in the constraint store. We now present reformulation techniques for a linear pseudo-Boolean inequality that *constrain* the reformulation of $cL \geq d$ w.r.t. S. In the previous sections we presented a method that reformulated a linear pseudo-Boolean inequality $cL \geq d$ as a set of extended clauses S' such that

$$\mathsf{Ext}(cL \geq d) = \mathsf{Ext}(S') \ .$$

Since $cL \geq d$ is part of a set of linear pseudo-Boolean inequalities S, it is sufficient if $cL \geq d$ is equivalent to S' w.r.t. the context S, i.e., only

$$\mathsf{Ext}(S \cup \{cL \geq d\}) = \mathsf{Ext}(S \cup S')$$

needs to hold. Assume that S' consists not only of extended clauses, i.e., arbitrary linear pseudo-Boolean inequalities are allowed. Then replacing $cL \geq d$ by S' is useful only if S' is in some sense *simpler* than $cL \geq d$.

We will present in Section 7.6.1 a method that generates a *stronger* linear pseudo-Boolean inequality $c'L' \geq d'$ from $cL \geq d$ with $\mathsf{Ext}(c'L' \geq d') \subseteq \mathsf{Ext}(cL \geq d)$ and

$$\mathsf{Ext}(S \cup \{cL \geq d\}) = \mathsf{Ext}(S \cup \{c'L' \geq d'\}) \ .$$

Thus, if $cL \geq d$ is part of the set of linear pseudo-Boolean inequalities S, we can replace $cL \geq d$ by the stronger linear pseudo-Boolean inequality $c'L' \geq d'$. It is often possible to transform some simple linear pseudo-Boolean inequalities in S into a small equivalent set of extended clauses, which then can be used to strengthen $cL \geq d$. Since extended clauses are computationally easier to handle, and nevertheless are the ultimate goal in our solved form, we restrict ourselves mainly to the case that the context S is a set of extended clauses. Hence, the goal is to generate a linear pseudo-Boolean inequality $c'L' \geq d'$ that dominates $cL \geq d$ with equivalence w.r.t. S being preserved, which means that $\mathsf{Ext}(S \cup \{cL \geq d\}) = \mathsf{Ext}(S \cup \{c'L' \geq d'\})$ still holds. Transforming such a strengthened linear pseudo-Boolean inequality $c'L' \geq d'$ instead of $cL \geq d$ generates stronger extended clauses.

In Section 7.6.2 we show how to replace $cL \geq d$ by a set of linear pseudo-Boolean inequalities S' in the context S such that $\mathsf{Ext}(S \cup \{cL \geq d\}) = \mathsf{Ext}(S \cup S')$. The linear pseudo-Boolean inequalities in S' are simpler in the sense that each of them contains fewer literals than $cL \geq d$. The method can be applied when the original linear pseudo-Boolean inequality $cL \geq d$ is too large to be transformed completely. We then hope to transform more easily some of the linear pseudo-Boolean inequalities in S' and afterwards strengthen the set of extended clauses, which again may lead to a strengthening of the remaining linear pseudo-Boolean inequalities in S'.

We will present a variant of the transformation method in Section 7.6.3. We again assume that the context of the linear pseudo-Boolean inequality $cL \geq d$ is a set of extended clauses S. We transform $cL \geq d$ into an equivalent set of extended clauses S' such that $\mathsf{Ext}(S \cup \{cL \geq d\}) = \mathsf{Ext}(S \cup S')$, but the number of extended clauses in S' is reduced or, optionally, the generated extended clauses are stronger than the one of the unconstrained transformation.

The *simplification* of a linear pseudo-Boolean inequality *constrained* by a context S is common to the methods presented in this section. In the context of constraint logic programming equivalence preserving reformulation of constraints into simpler ones in a context is called *relative simplification* [Smo94]; an essential functionality of constraint solvers for concurrent constraint languages.

7.6.1 Coefficient Reduction

Let S be a set of linear pseudo-Boolean inequalities, the context, and $cL \geq d$ be the linear pseudo-Boolean inequality that we want to replace. We will present a method that reduces a coefficient of $cL \geq d$ such that the modified linear pseudo-Boolean inequality dominates $cL \geq d$, but is still equivalent to $cL \geq d$ in the context S.

Let c_iL_i be a product of $cL \geq d$ and let $c'L'$ be $cL \setminus \{c_iL_i\}$. We search for a linear pseudo-Boolean inequality $c'_iL_i + c'L' \geq d$ with $c'_i < c_i$, which implies that $c'_iL_i + c'L' \geq d$ dominates $cL \geq d$, and

$$\mathsf{Ext}(S \cup \{c'_iL_i + c'L' \geq d\}) = \mathsf{Ext}(S \cup \{cL \geq d\}) \ . \tag{7.20}$$

Since $c'_i L_i + c' L' \geq d$ dominates $cL \geq d$, we know that $\mathsf{SCP}(\mathsf{SRed}(c'_i L_i + c' L' \geq d))$ dominates $\mathsf{SCP}(\mathsf{SRed}(cL \geq d))$. Because our transformation method generates a set of prime extended clauses, we further know that each extended clause in $\mathsf{SCP}(\mathsf{SRed}(cL \geq d))$ is dominated by an extended clause in $\mathsf{SCP}(\mathsf{SRed}(c'_i L_i + c' L' \geq d))$. Transforming $c'_i L_i + c' L' \geq d$ instead of $cL \geq d$ generates therefore only stronger extended clauses, since every extended clause we obtain by transforming $cL \geq d$ is dominated by an extended clause obtained by transforming $c'_i L_i + c' L' \geq d$.

Next, we describe how to derive a coefficient c'_i such that (7.20) holds. The key idea is to deduce a lower bound b for the linear pseudo-Boolean term $c' L'$ w.r.t. $S \cup \{L_i \geq 1\}$.

Proposition 7.6.1 *If $S \cup \{L_i \geq 1\}$ dominates $c' L' \geq b$ and $c_i > d - b$ then*

$$\mathsf{Ext}(S \cup \{c_i L_i + c' L' \geq d\}) = \mathsf{Ext}(S \cup \{(d-b) L_i + c' L' \geq d\}) \ .$$

We then have reduced the coefficient c_i to $c'_i = d - b < c_i$.

Proof. The direction $\supseteq$ is obvious, since $c_i > d - b$ and thus $(d-b)L_i + c' L' \geq d$ dominates $c_i L_i + c' L' \geq d$. It remains to show $\subseteq$. Let α be in $\mathsf{Ext}(S \cup \{c_i L_i + c' L' \geq d\})$. We show that then α is in $\mathsf{Ext}(S \cup \{(d-b) L_i + c' L' \geq d\})$ by case analysis on $\alpha(L_i)$.

$\alpha(L_i) = 0$: Then $\alpha(c_i L_i + c' L') = \alpha((d-b) L_i + c' L') = \alpha(c' L')$, and thus we know that $\alpha(c_i L_i + c' L') \geq d$ if and only if $\alpha((d-b) L_i + c' L') \geq d$.

$\alpha(L_i) = 1$: Then $\alpha((d-b) L_i + c' L') = (d-b) + \alpha(c' L')$. It remains to show that $(d-b) + \alpha(c' L') \geq d$, or $\alpha(c' L') \geq b$, which follows from the hypothesis that $S \cup \{L_i \geq 1\}$ dominates $c' L' \geq b$. □

Dietrich et al. [DEC93] present a similar coefficient reduction technique, where essentially the bounds are obtained in a different way. A necessary condition for the application of coefficient reduction is that $S \cup \{L_i \geq 1\}$ dominates $cL \geq d$. Otherwise S does not dominate $c' L' \geq d - c_i$ and therefore for each lower bound b for $c' L'$ w.r.t. $S \cup \{L_i \geq 1\}$ we have $b \leq d - c_i$, which contradicts $c_i > d - b$.

Example 7.6.2 Let $cL \geq d$ be

$$10 \cdot A + 5 \cdot B + 4 \cdot C + 3 \cdot D + 3 \cdot E \geq 10 \ . \tag{7.21}$$

Good candidates for coefficient reductions are obviously coefficient/literal pairs $c_i L_i$, where the coefficient c_i is relatively large w.r.t. the right-hand side, because then the necessary condition, $S \cup \{L_i \geq 1\}$ dominates $cL \geq d$, is more likely to hold. Let us choose $10 \cdot A$ and, since the coefficient 10 equals the right-hand side, we know that $S \cup \{A \geq 1\}$ dominates (7.21) for all sets of extended clauses S. We can reduce the coefficient 10 if we find some lower bound $b > 0$ for the linear pseudo-Boolean term $5 \cdot B + 4 \cdot C + 3 \cdot D + 3 \cdot E$. Suppose that S contains the extended clause $\overline{A} + B + C + F \geq 2$. Since $\{\overline{A} + B + C + F \geq 2, A \geq 1\}$ dominates $B + C \geq 1$, we know that either B or C has to be 1. Therefore, we can derive the lower bound $b = 4$.

$$\{\overline{A} + B + C + F \geq 2, A \geq 1\} \text{ dominates } 5 \cdot B + 4 \cdot C + 3 \cdot D + 3 \cdot E \geq 4$$

Following Proposition 7.6.1 we obtain $c_i' = 6$ and $c_i'L_i + cL \geq d$ is

$$6 \cdot A + 5 \cdot B + 4 \cdot C + 3 \cdot D + 3 \cdot E \geq 10 \ .$$

When transforming both linear pseudo-Boolean inequalities $c_iL_i + c'L' \geq d$ and $c_i'L_i + c'L' \geq d$ into an equivalent set of extended clauses, we see that the former generates only extended clauses with right-hand side 1, whereas the latter generates for example the strongest extended clause $A + B + C + D \geq 2$.

In order to obtain an ideal coefficient reduction we need to derive the maximal lower bound $b_{\min}$ for $c'L'$ w.r.t. $S \cup \{L_i \geq 1\}$. Because this involves solving the NP-complete problem

$$\min : c'L' \qquad \text{w.r.t.} \quad S \cup \{L_i \geq 1\} \ ,$$

we only approximate the best lower bound $b_{\min}$.

We propose the following procedure to obtain a valid lower bound b. Since the constraint set contains the unit literal L_i, we first simplify the set of extended clauses by replacing $S \cup \{L_i \geq 1\}$ by $\texttt{pbur}(S \cup \{L_i \geq 1\})$. Note that $S \cup \{L_i \geq 1\}$ dominates $\texttt{pbur}(S \cup \{L_i \geq 1\})$. We then fix the literals of $\texttt{pbul}(S \cup \{L_i \geq 1\})$ in the objective function. For a set of literals K we define

$$\texttt{o\_reduce}(c'L', K) := \{c_jL_j \mid c_jL_j \in c'L', L_j \notin K \text{ and } \overline{L_j} \notin K\}$$

and the new objective function is $\hat{c}\hat{L} := \texttt{o\_reduce}(c'L', \texttt{pbul}(S \cup \{L_i \geq 1\}))$. Summing up the coefficients of the literals that are fixed to 1 yields the constant $\hat{b}$, which can be added to the lower bound that we obtain from the approximation of the minimum for $\hat{c}\hat{L}$. For a set of literals K we define

$$\texttt{sum\_fixed}(c'L', K) := \sum_{c_jL_j \in c'L' : L_j \in K} c_j$$

and the desired constant is then

$$\hat{b} := \texttt{sum\_fixed}(c'L', \texttt{pbul}(S \cup \{L_i \geq 1\})) \ .$$

The minimum of $c'L'$ w.r.t. $S \cup \{L_i \geq 1\}$ is then the same as the minimum of $\hat{c}\hat{L} + \hat{b}$ w.r.t. $\texttt{pbur}(S \cup \{L_i \geq 1\})$.

We restrict ourselves to a context S consisting of extended clauses and consider the problem of approximating the minimum of $\hat{c}\hat{L}$ w.r.t. the set of extended clauses $\texttt{pbur}(S \cup \{L_i \geq 1\})$, where no extended clause in $\texttt{pbur}(S \cup \{L_i \geq 1\})$ dominates an extended clause of the form $L_i \geq 1$. In other words, no more obvious fixings apply. Next, we will derive a set of extended clauses S' from the set $\texttt{pbur}(S \cup \{L_i \geq 1\})$ such that S' is dominated by $\texttt{pbur}(S \cup \{L_i \geq 1\})$ and S' contains only literals also occurring in $\hat{L}$. For a set of literals K and a set of extended clauses T we define

$$\texttt{c\_reduce}(T, K) := \left\{ K' \geq \beta' \;\middle|\; \begin{array}{l} K' \uplus K'' \geq \beta \in T, \\ \beta' = \beta - |K''| \geq 1, \\ K'' \subseteq K \text{ and } K \cap K' = \emptyset \end{array} \right\} ,$$

which is the set of all strict reductions of the extended clauses in T, where all literals in K have been eliminated. The set of extended clauses is then

$$S' = \texttt{c\_reduce}(\texttt{pbur}(S \cup \{L_i \geq 1\}), \mathcal{L} \setminus (\hat{L} \cup \overline{\hat{L}})) ,$$

where $\mathcal{L}$ is the set of all literals. We now give a greedy like approach to obtain a valid lower bound b for $\hat{c}\hat{L}$ w.r.t. the set of extended clauses S'. For each extended clause $K' \geq \beta'$ in S' we define

$$b_{K' \geq \beta'} := s(\hat{c}') \qquad \text{where} \quad \begin{cases} \hat{c}'\hat{L}' \subseteq \hat{c}\hat{L}, \\ \hat{L}' \subseteq K', \\ |\hat{L}'| = \beta' \text{ and} \\ \max(\hat{c}') \leq \min(\hat{c} \setminus \hat{c}') \ . \end{cases}$$

That is, $b_{K' \geq \beta'}$ is the sum of the smallest β' coefficients of $\hat{c}\hat{L}$ whose literals are in K'. So for each $\alpha \in \mathsf{Ext}(K' \geq \beta')$ we have $\alpha(\hat{c}\hat{L}) \geq b_{K' \geq \beta'}$ and so a first valid lower bound for $\hat{c}\hat{L}$ is $b_{K' \geq \beta'}$. Let us select some bound $b_{K' \geq \beta'}$. For deriving the bound $b_{K' \geq \beta'}$ we considered all literals occurring in K'. Note that the bound is also valid for the subterm of the objective function containing only the literals of K'. We can therefore safely replace this subterm in the objective function by the bound $b_{K' \geq \beta'}$ and minimize the simpler objective function. Hence, a valid lower bound for $\hat{c}\hat{L}$ can be obtained by solving a simpler minimization problem

$$\min : \texttt{o\_reduce}(\hat{c}\hat{L}, K') + b_{K' \geq \beta'} \qquad \text{subject to} \quad S' \ ,$$

where $K' \geq \beta' \in S'$, which can again be solved by the same approach. We obtain the following algorithm computing a valid lower bound b for $\hat{c}\hat{L}$ w.r.t. S'.

Algorithm 7.6.3 *Lower bound for a linear pseudo-Boolean term*

```
lower_bound(ĉL̂, S')
        b := 0
        while ĉL̂ ≠ ∅ ∧ S' ≠ ∅
                K' ≥ β' := select_clause(ĉL̂, S')
                b := b + b_{K'≥β'}
                S' := c_reduce(S', K')
                ĉL̂ := o_reduce(ĉL̂, K')
        endwhile
        return b
end lower_bound
```

The procedure `select_clause` selects an extended clause that is used for deriving the current bound. We use the following heuristics. We choose the extended clause $K' \geq \beta'$ that maximizes

$$\frac{b_{K' \geq \beta'}}{\sum_{c_i L_i \in \hat{c}\hat{L} : L_i \in K'} c_i} .$$

The heuristics reflects that the bound is the largest or best w.r.t. the sum of the coefficients it has consumed. The computed lower bound b for $c'L'$ w.r.t. $S \cup \{L_i \geq 1\}$ is then

`sum_fixed`$(c'L', K)+$
`lower_bound`$($`o_reduce`$(c'L', K),$ `c_reduce`$($**pbur**$(S \cup \{L_i \geq 1\}), L' \setminus K))$

with $K =$ **pbul**$(S \cup \{L_i \geq 1\})$.

Example 7.6.4 Let $cL \geq d$ be

$$5 \cdot A + 4 \cdot B + 4 \cdot C + 4 \cdot D + 3 \cdot E + 2 \cdot F \geq 11 \tag{7.22}$$

and let the set of extended clauses S be

$$\{\overline{A} + B \geq 1, C + D + G \geq 2, D + E + F \geq 1\} .$$

Let us reduce the coefficient 5 of the literal A. We apply pseudo-Boolean unit resolution to $S \cup \{A \geq 1\}$ and obtain

$$\begin{aligned} \texttt{pbur}(S \cup \{A \geq 1\}) &= \{C + D + G \geq 2, D + E + F \geq 1\} \text{ and} \\ \texttt{pbul}(S \cup \{A \geq 1\}) &= \{A, B\} \ . \end{aligned}$$

Next, we reduce $\texttt{pbur}(S \cup \{A \geq 1\})$ such that only the literals of the objective function remain, i.e., we eliminate all literals in $\mathcal{L} \setminus \{C, D, E, F, \overline{C}, \overline{D}, \overline{E}, \overline{F}\}$. and obtain

$$S' = \{C + D \geq 1, D + E + F \geq 1\} \ .$$

We update the objective function $c'L'$ by calculating

$$\begin{aligned} \texttt{sum\_fixed}(4 \cdot B + 4 \cdot C + 4 \cdot D + 3 \cdot E + 2 \cdot F, \{A, B\}) = 4 \\ \texttt{o\_reduce}(cL, \{A, B\}) = 4 \cdot C + 4 \cdot D + 3 \cdot E + 2 \cdot F = \hat{c}\hat{L} \end{aligned}$$

and then call `lower_bound`. Calculating the $b_{K' \geq \beta'}$ yields

$$b_{C+D \geq 1} = 4 \qquad b_{D+E+F \geq 1} = 2$$

and since $4/8 > 2/9$, we select $C + D \geq 1$. Updating the objective function and the extended clause set gives

$$\begin{aligned} \texttt{o\_reduce}(4 \cdot C + 4 \cdot D + 3 \cdot E + 2 \cdot F, \{C, D\}) &= 3 \cdot E + 2 \cdot F \\ \texttt{c\_reduce}(\{C + D \geq 1, D + E + F \geq 1\}, \{C, D\}) &= \emptyset \end{aligned}$$

and hence `lower_bound` stops and returns 4. The overall bound b is then $4 + 4 = 8$ and therefore $S \cup \{A \geq 1\}$ dominates

$$4 \cdot B + 4 \cdot C + 4 \cdot D + 3 \cdot E + 2 \cdot F \geq 8 \ .$$

By Proposition 7.6.1 we have $5 = c_i > d - b = 11 - 8 = 3$ and can so reduce the coefficient 5 of A to 3. Finally, we obtain the stronger linear pseudo-Boolean inequality

$$4 \cdot B + 4 \cdot C + 4 \cdot D + 3 \cdot A + 3 \cdot E + 2 \cdot F \geq 11 \ ,$$

which is equivalent to (7.22) in the context S.

The presented method for deriving a lower bound for $c'L'$ w.r.t. $S \cup \{L_i \geq 1\}$ yields good bounds when the set of extended clauses S is sparse or if the set of extended clauses S is nearly prime. Note that any method can be used to obtain a valid lower bound and the better the approximation of $b_{\min}$ the more likely it is that coefficient reduction applies.

The coefficient reduction method is useful for strengthening 0-1 problems that contain so-called *big-M-constraints* of the form

$$M \cdot X_i + cL \geq d$$

with $M = d$. We call X_i a *decision variable*, since setting it to 0 activates the constraint $cL \geq d$, whereas setting it to 1 makes $M \cdot X_i + cL \geq d$ a tautology [Wil90]. For these linear pseudo-Boolean inequalities all generated extended clauses obtained by the transformation process have the right-hand side 1. But, it is likely that coefficient reduction applies, since even a lower bound 1 for cL reduces M by 1. Note that then the transformation also generates stronger extended clauses with a right-hand side greater than 1.

7.6.2 Constrained Strict Reductions

Next, we introduce a method that replaces a linear pseudo-Boolean inequality $cL \geq d$ by a set of linear pseudo-Boolean inequalities S' such that $cL \geq d$ is equivalent to S' w.r.t. a set of extended clauses S.

$$\mathsf{Ext}(S \cup \{cL \geq d\}) = \mathsf{Ext}(S \cup S')$$

In the previous section we focused on deriving a lower bound for a linear pseudo-Boolean subterm of the original linear pseudo-Boolean inequality. With help of this lower bound we strengthened the linear pseudo-Boolean inequality and therefore the extended clauses generated by the transformation method. Application of coefficient reduction is always an improvement. In contrast to the method of Section 7.6.1 we allow here that S' is weaker than $cL \geq d$, which means $\mathsf{Ext}(S') \supseteq \mathsf{Ext}(cL \geq d)$. The set S' is simpler because the linear pseudo-Boolean inequalities in S' contain fewer literals than $cL \geq d$. Unfortunately, we may lose strong extended clauses when transforming the linear pseudo-Boolean inequalities in S' instead of $cL \geq d$. On the other hand, we possibly reduce the number of generated extended clauses. The idea is that some of the linear pseudo-Boolean inequalities in S' are easier to transform than $cL \geq d$. Combining the obtained strongest extended clauses with S may strengthen S, which again may lead to further simplifications.

When transforming a linear pseudo-Boolean inequality $cL \geq d$ we need to calculate all strict reductions $\mathsf{SRed}(cL \geq d)$. We show how to *constrain* the generation of strict reductions, and thus reduce their number with equivalence still being preserved.

Proposition 7.6.5 *Let $cL \geq d$ be a linear pseudo-Boolean inequality and let $L' \geq \beta$ be an extended clause with $L' \subseteq L$. Let S' be the set of linear pseudo-Boolean inequalities*

$$\left\{ \hat{c}\hat{L} \geq \hat{d} \;\middle|\; \begin{array}{l} \hat{c}\hat{L} = cL \setminus c''L'', \hat{d} = d - \sum c'', \\ c''L'' \subseteq cL, L'' \subseteq L' \text{ and } |L''| = \beta \end{array} \right\},$$

then $\mathsf{Ext}(\{L' \geq \beta, cL \geq d\}) = \mathsf{Ext}(\{L' \geq \beta\} \cup S')$. *Note that S' is the set of all strict reductions of $cL \geq d$, where the set of literals L'' for all $L'' \subseteq L'$ with $|L''| = \beta$ has been eliminated.*

Proof. Since $S' \subseteq \mathsf{SRed}(cL \geq d)$, we know that $cL \geq d$ dominates S'. It remains to show that every solution α of $\{L' \geq \beta\} \cup S'$ is a solution of $cL \geq d$. Let α be in $\mathsf{Ext}(\{L' \geq \beta\} \cup S')$. Let $L''' \subseteq L'$ be the set of all literals $L_i \in L'$ with $\alpha(L_i) = 1$. Because $L' \geq \beta$, we know that $|L'''| \geq \beta$. Let L'' be a subset of L''' such that $|L''| = \beta$. We select the strict reduction $\hat{c}\hat{L} \geq d - \sum c''$ from S', where L'' has been eliminated. Because α is a solution of S', we know that $\alpha(\hat{c}\hat{L}) \geq d - \sum c''$ and therefore $\alpha(\hat{c}\hat{L}) + \sum c'' \geq d$. Because $\alpha(\hat{c}\hat{L}) + \sum c'' = \alpha(cL)$, we have $\alpha(cL) \geq d$. □

Example 7.6.6 Let $cL \geq d$ be a linear pseudo-Boolean inequality and let $\{c_i \cdot A, c_j \cdot B, c_k \cdot C\} \subseteq cL$. We suppose that the extended clause $A + B + C \geq 2$ is dominated by an extended clause in the context S. Note that c_i, c_j, c_k can be any coefficients of $cL \geq d$, not necessarily the largest ones. Let $\hat{c}\hat{L} = cL \setminus \{c_i \cdot A, c_j \cdot B, c_k \cdot C\}$. We then can replace $cL \geq d$ by

$$\begin{aligned} S' = \{c_i \cdot A + \hat{c}\hat{L} &\geq d - (c_j + c_k), \\ c_j \cdot B + \hat{c}\hat{L} &\geq d - (c_i + c_k), \\ c_k \cdot C + \hat{c}\hat{L} &\geq d - (c_i + c_j)\} \end{aligned}$$

according to Proposition 7.6.5, where $L' \geq \beta$ is $A + B + C \geq 2$, and have $\mathsf{Ext}(S \cup \{cL \geq d\}) = \mathsf{Ext}(S \cup S')$.

The use of Proposition 7.6.5 is only practicable if the number of linear pseudo-Boolean inequalities it generates is not too high. Note that Proposition 7.6.5 generates $\binom{|L'|}{\beta}$ extended clauses. The most interesting cases are if β is close to $|L'|$, since then the number of linear pseudo-Boolean inequalities is small and the number of deleted literals is high. In Example 7.6.6 we used an extended clause with $|L'| - \beta = 1$ and so $\binom{|L'|}{\beta} = |L'|$. Extended clauses of the form $L' \geq |L'| - 1$ are also called *clique inequalities* or *special ordered set constraints* [NW88] and frequently occur in practical linear pseudo-Boolean constraint sets.

One problem of the application of Proposition 7.6.5 is that we may lose strong extended clauses.

Example 7.6.7 [a] Let $cL \geq d$ be

$$5 \cdot A + 4 \cdot B + 3 \cdot C \geq 6$$

then $cL \geq d$ is equivalent to $A + B + C \geq 2$. Applying Proposition 7.6.5 on $cL \geq d$ with the extended clause $A + B \geq 1$ gives the set of linear pseudo-Boolean inequalities

$$\{5 \cdot A + 3 \cdot C \geq 2, 4 \cdot B + 3 \cdot C \geq 1\} \tag{7.23}$$

equivalent to $cL \geq d$ w.r.t. $\{A + B \geq 1\}$. Transforming the linear pseudo-Boolean inequalities in (7.23) yields

$$\{A + C \geq 1, B + C \geq 1\} \ .$$

Now Proposition 7.6.5 assures that

$$\mathsf{Ext}(\{A + C \geq 1, B + C \geq 1, A + B \geq 1\}) = \mathsf{Ext}(\{A + B + C \geq 2\}) \ ,$$

but the *better* formulation of $cL \geq d$ is obviously $\{A + B + C \geq 2\}$. Indeed, in Section 8.2 we describe how to obtain this better formulation if only $\{A + C \geq 1, B + C \geq 1\}$ and $A + B \geq 1$ are given.

[b] Let $cL \geq d$ be

$$d \cdot A + c_i \cdot B + \hat{c}\hat{L} \geq d \ ,$$

where $c_i \cdot B$ can be any coefficient literal pair in $cL \setminus \{d \cdot A\}$. Then we know that all generated extended clauses have degree 1, since there is a coefficient in $cL \geq d$ that equals the right-hand side. Note that there is no strict reduction on A, because eliminating A would give a tautology. Suppose that there is a clique inequality in S dominating $A + B \geq 1$. Because of Proposition 7.6.5, we know that $cL \geq d$ is equivalent to

$$(d - c_i) \cdot A + \hat{c}\hat{L} \geq d - c_i$$

and we are sure that we do not miss any strongest extended clause. So applications of Proposition 7.6.5, where all but one[1] of the generated strict reductions are dominated by the context are safe.

[1] Of course, if all generated strict reductions are dominated by the context, then the whole linear pseudo-Boolean inequality is dominated by the context and can be ignored.

Another problem is that transforming the linear pseudo-Boolean inequalities in the set S' of Proposition 7.6.5 may invoke redundant computation. While transforming the linear pseudo-Boolean inequalities of S' in Example 7.6.6 we fully transform 3 times the strict reduction $cL \geq d - (c_i + c_j + c_k)$. We avoid the redundant computation by using Proposition 7.6.5 directly in the transformation routine.

Example 7.6.8 Let $cL \geq d$ be a linear pseudo-Boolean inequality and let $\{c_i \cdot A, c_j \cdot B, c_k \cdot C\} \subseteq cL$ as in Example 7.6.6. Let $\hat{c}\hat{L} = cL \backslash \{c_i \cdot A, c_j \cdot B, c_k \cdot C\}$ and $c_i \geq c_j \geq c_k$. We assume that the extended clause $A + B + C \geq 2$ is dominated by an extended clause in the context. Because of Proposition 7.6.5, we know that $cL \geq d$ is equivalent to the set of linear pseudo-Boolean inequalities

$$\begin{aligned} S' = \{c_i \cdot A + \hat{c}\hat{L} &\geq d - (c_j + c_k), \\ c_j \cdot B + \hat{c}\hat{L} &\geq d - (c_i + c_k), \\ c_k \cdot C + \hat{c}\hat{L} &\geq d - (c_i + c_j)\} \ . \end{aligned}$$

We transform $c_i \cdot A + \hat{c}\hat{L} \geq d - (c_j + c_k)$ completely by the standard transformation method, but we start the transformation of $c_j \cdot B + \hat{c}\hat{L} \geq d - (c_i + c_k)$ with the forbidden elimination on B, since all extended clauses generated when B is eliminated are already obtained by the full transformation of $c_i \cdot A + \hat{c}\hat{L} \geq d - (c_j + c_k)$. Likewise, we forbid elimination of C when transforming $c_k \cdot C + \hat{c}\hat{L} \geq d - (c_i + c_j)$ and so avoid the redundant transformation of $\hat{c}\hat{L} \geq d - (c_i + c_j + c_k)$.

7.6.3 Constrained Transformation

Finally, we present a variant of the transformation method that restricts the number of generated extended clauses while transforming a linear pseudo-Boolean inequality $cL \geq d$ w.r.t. a set of extended clauses S. Let E be the set of non-redundant extended clauses generated by the transformation of $cL \geq d$, then *constrained transformation* generates a smaller set $E' \subseteq E$ such that

$$\mathsf{Ext}(S \cup \{cL \geq d\}) = \mathsf{Ext}(S \cup E') \ .$$

While transforming a linear pseudo-Boolean inequality $cL \geq d$ we need to compute all strict reductions $\mathsf{SRed}(cL \geq d)$. When proving the equivalence of $\mathsf{SCP}(\mathsf{SRed}(cL \geq d))$ and $cL \geq d$ in Theorem 7.1.5 we showed that for each non-satisfying assignment α of $cL \geq d$, we find an extended clause $L' \geq \beta$ in $\mathsf{SCP}(\mathsf{SRed}(I))$ such that α is not a solution of $L' \geq \beta$. If α is not a solution of the context S, then we need not find an extended clause in $\mathsf{SCP}(\mathsf{SRed}(cL \geq d))$ for which α is not a solution. Hence, the corresponding generated extended clause need not be added for preserving equivalence. We can so restrict the number of needed strict reductions and therefore the number of generated extended clauses. First, we define a set of strict reductions of $cL \geq d$ taking into account the context S.

Definition 7.6.9 Let $cL \geq d$ be a linear pseudo-Boolean inequality, let S be a set of extended clauses and let $c'L' \geq d'$ be a strict reduction of $cL \geq d$. The set of literals $L \setminus L'$ is the set of literals that have been eliminated from $cL \geq d$ in order to obtain $c'L' \geq d'$. We denote by $\mathsf{SRed}(cL \geq d, S) \subseteq \mathsf{SRed}(cL \geq d)$ the set of strict reductions of $cL \geq d$ such that

$$\begin{array}{rl} & c'L' \geq d' \in \mathsf{SRed}(cL \geq d, S) \\ \text{if and only if} & S \cup \{L \setminus L' \geq |L \setminus L'|\} \text{ is satisfiable.} \end{array}$$

We now show that the set $\mathsf{SCP}(\mathsf{SRed}(cL \geq d, S))$ of strongest extended clauses obtained from the smaller set of strict reductions $\mathsf{SRed}(cL \geq d, S)$ is sufficient for the equivalence with $cL \geq d$ w.r.t. a set S of linear pseudo-Boolean inequalities.

Theorem 7.6.10 *Let $cL \geq d$ be a linear pseudo-Boolean inequality and let S be a set of extended clauses, then*

$$\mathsf{Ext}(S \cup \{cL \geq d\}) = \mathsf{Ext}(S \cup \mathsf{SCP}(\mathsf{SRed}(cL \geq d, S)s)) \ .$$

Proof.

- $\mathsf{Ext}(S \cup \{cL \geq d\}) \subseteq \mathsf{Ext}(S \cup \mathsf{SCP}(\mathsf{SRed}(cL \geq d, S)))$: as in Theorem 7.1.5.
- $\mathsf{Ext}(S \cup \{cL \geq d\}) \supseteq \mathsf{Ext}(S \cup \mathsf{SCP}(\mathsf{SRed}(cL \geq d, S)))$:
  We show that $\alpha \notin \mathsf{Ext}(S \cup \mathsf{SCP}(\mathsf{SRed}(cL \geq d, S)))$ if $\alpha \notin \mathsf{Ext}(S \cup \{cL \geq d\})$ from which the theorem follows.
  Suppose that $\alpha \notin \mathsf{Ext}(S \cup \{cL \geq d\})$. If $\alpha \notin \mathsf{Ext}(S)$ then obviously $\alpha \notin \mathsf{Ext}(S \cup \mathsf{SCP}(\mathsf{SRed}(cL \geq d, S)))$.
  Suppose that $\alpha \in \mathsf{Ext}(S)$, but $\alpha \notin \mathsf{Ext}(cL \geq d)$. We show that in this case there is a strict reduction in $\mathsf{SRed}(cL \geq d, S)$ such that α is not a solution of this reduction, hence $\alpha \notin \mathsf{Ext}(\mathsf{SCP}(\mathsf{SRed}(cL \geq d, S)))$ and thus $\alpha \notin \mathsf{Ext}(S \cup \mathsf{SCP}(\mathsf{SRed}(cL \geq d, S)))$. Since $\alpha \notin \mathsf{Ext}(cL \geq d)$, we know that $\alpha(cL) = d' < d$. Let Y be the set of literals L_i such that $\alpha(L_i) = 1$ and $L' = L \setminus Y$, then $\sum_{i|L_i \in Y} c_i = d' < d$ and $c'L' \geq d'$ is

$\sum_{i|L_i \in L'} c_i L_i \geq d - d'$. Because $d' < d$, we know that $d - d' \geq 1$, and therefore $c'L' \geq d'$ is a strict reduction of $cL \geq d$, obtained by eliminating the literals that are mapped to zero by α. Furthermore, we know that $Y = L \setminus L'$, hence $\alpha(L \setminus L') = |L \setminus L'|$. Because $\alpha \in \mathsf{Ext}(S)$, we know that $\alpha \in \mathsf{Ext}(S \cup \{L \setminus L' \geq |L \setminus L'|\})$ and therefore $c'L' \geq d'$ in $\mathsf{SRed}(cL \geq d, S)$. Let $\mathsf{SCP}(c'L' \geq d')$ be $L' \geq \beta$. We know that $\beta \geq 1$ because $d - d' \geq 1$. Because $\alpha(L') = 0$, we derive that $\alpha \notin \mathsf{Ext}(\mathsf{SCP}(c'L' \geq d'))$, and therefore $\alpha \notin \mathsf{Ext}(\mathsf{SCP}(\mathsf{SRed}(cL \geq d)))$, which establishes the theorem. □

The proof of Theorem 7.6.10 is similar to the proof of Theorem 7.1.5. The main difference is that we additionally show that each strict reduction we need is in the restricted set of strict reductions. Theorem 7.1.5 can be seen as an instance of Theorem 7.6.10, where the set of extended clauses S is the empty set.

An implementation of the constrained transformation algorithm is now straightforward. For each strict reduction $c'L' \geq d'$ of $cL \geq d$ we maintain a set of extended clauses $S_{c'L' \geq d'} := S \cup \{L \setminus L' \geq |L \setminus L'|\}$. By Theorem 7.6.10 we can eliminate the strict reduction $c'L' \geq d'$ from the transformation process if $S_{c'L' \geq d'}$ is unsatisfiable. Note that if $S_{c'L' \geq d'}$ is unsatisfiable, then for each strict reduction $c''L'' \geq d''$ of $c'L' \geq d'$ the corresponding set of extended clauses $\{S \cup \{L \setminus L'' \geq |L \setminus L''|\}$ is unsatisfiable. Care must be taken when applying the strong redundancy criteria presented in Section 7.2. Suppose that a strict reduction $c'L' \geq d'$ is eliminated because $S_{c'L' \geq d'}$ is unsatisfiable. Furthermore, suppose that $\mathsf{SCP}(c'L' \geq d')$ dominates an extended clause $\mathsf{SCP}(c''L'' \geq d'')$, where $c''L'' \geq d''$ is a strict reduction of $c'L' \geq d'$ and there is no other non-redundant extended clause dominating $\mathsf{SCP}(c'L' \geq d')$. In that case we have to add $\mathsf{SCP}(c''L'' \geq d'')$ or $\mathsf{SCP}(c'L' \geq d')$ for preserving equivalence. The simplest way is to add $\mathsf{SCP}(c'L' \geq d')$ always, if $S_{c'L' \geq d'}$ is unsatisfiable, and then forbid further strict reductions of $c'L' \geq d'$, i.e., not to transform recursively $c'L' \geq d'$. When later inserting the transformed set of extended clauses into a larger set, redundant extended clauses can be deleted.

In the constrained transformation algorithm we need to solve again a satisfiability problem for the set of extended clauses $S \cup \{L \setminus L' \geq |L \setminus L'|\}$. Because this is an NP-complete problem, we relax the problem again and check only whether the pseudo-Boolean unit relaxation of $S \cup \{L \setminus L' \geq |L \setminus L'|\}$ is unsatisfiable. That is, at each node we check whether $\perp \in \mathtt{pbur}(S \cup \{L \setminus L' \geq |L \setminus L'|\})$. Note that all literals in $L \setminus L'$ are unit literals and therefore unsatisfiability of $S \cup \{L \setminus L' \geq |L \setminus L'|\}$ is likely to be detected by pseudo-Boolean unit resolution.

Example 7.6.11 Let $cL \geq d$ be

$$6 \cdot A + 5 \cdot B + 4 \cdot C + 3 \cdot D + 2 \cdot E + F \geq 13$$

and suppose that the extended clause $\overline{B} + \overline{C} \geq 1$ is dominated by an extended clause in the context S. Hence, $S \cup \{B + C \geq 2\}$ is unsatisfiable which is detected by pseudo-Boolean unit resolution. The full transformation of $cL \geq d$ generates the five extended clauses

$$\{ \quad A + B + C + D + E \geq 3, A + B + C \geq 2, A + D \geq 1, \\ A + B + D + F \geq 2, A + E + F \geq 1\} \ .$$

One strict reduction of $cL \geq d$ is

$$6 \cdot A + 3 \cdot D + 2 \cdot E + F \geq 4 \tag{7.24}$$

from which the two non-redundant extended clauses

$$A + E + F \geq 1 \qquad A + D \geq 1$$

are derived. Since the set of extended clauses $S_{6 \cdot A + 3 \cdot D + 2 \cdot E + F \geq 4}$ is unsatisfiable, and therefore $\mathsf{SRed}(6 \cdot A + 3 \cdot D + 2 \cdot E + F \geq 4, S) = \emptyset$, we need not consider $6 \cdot A + 3 \cdot D + 2 \cdot E + F \geq 4$ and constrained transformation generates only three extended clauses. Note that $\mathsf{SCP}(6 \cdot A + 3 \cdot D + 2 \cdot E + F \geq 4)$ is redundant and dominated by another extended clause from the remaining three. Hence, we need not include $\mathsf{SCP}(6 \cdot A + 3 \cdot D + 2 \cdot E + F \geq 4)$.

Instead of deleting the strict reduction $c'L' \geq d'$ if $S_{c'L' \geq d'}$ is unsatisfiable, we can also use this fact for strengthening the generated extended clause. We explain the idea first by an example. Let $cL \geq d$ be

$$c_1 \cdot A + c_2 \cdot B + cL \geq d$$

and suppose that $S \cup \{A + B \geq 2\}$ is unsatisfiable. Hence, the strict reduction

$$cL \geq d - (c_1 + c_2) \ . \tag{7.25}$$

need not be considered. However, because $S \cup \{A + B \geq 2\}$ is unsatisfiable we know that $\overline{A} + \overline{B} \geq 1$ is a valid extended clause w.r.t. S. So we can derive an upper bound for the linear pseudo-Boolean term $c_1 \cdot A + c_2 \cdot B$, namely $\{\overline{A} + \overline{B} \geq 1\}$ dominates $c_1 \cdot A + c_2 \cdot B \leq \max(c_1, c_2)$. Therefore, we derive the stronger linear pseudo-Boolean inequality

$$cL \geq d - \max(c_1, c_2) \ . \tag{7.26}$$

We miss no extended clause if we replace the strict reduction (7.25) by (7.26), but may only generate stronger extended clauses. Note that we derive here an *upper bound* for a linear pseudo-Boolean term and use the bound in order to strengthen the generated extended clauses.

Lemma 7.6.12 *Let $cL \geq d$ be a linear pseudo-Boolean inequality and let $c'L' \geq d'$ be a strict reduction of $cL \geq d$. Let S be a set of extended clauses such that $S_{c'L' \geq d'}$ is unsatisfiable. Let $c''L''$ be $cL \setminus c'L'$, then*

$$\mathsf{Ext}(S \cup \{cL \geq d\}) = \mathsf{Ext}(S \cup \mathsf{SCP}(\mathsf{SRed}(cL \geq d, S)) \cup \{c'L' \geq d' + \min(c'')\}) \ .$$

Proof. The direction $\supseteq$ is immediate by Theorem 7.6.10. It remains to show $\subseteq$, which means $S \cup \{cL \geq d\}$ dominates $c'L' \geq d' + \min(c'')$. Let α be in $\mathsf{Ext}(S \cup \{cL \geq d\})$, then we have $\alpha(cL) \geq d$ and therefore $\alpha(c'L') + \alpha(c''L'') \geq d$. Since $S_{c'L' \geq d'} = S \cup \{L'' \geq |L''|\}$ is unsatisfiable, we know furthermore that S dominates $\overline{L''} \geq 1$, i.e., at least one literal of L'' is mapped to 0 by α. We conclude that $\alpha(c''L'') \leq \sum c'' - \min(c'')$ and, since $\alpha(c'L') + \alpha(c''L'') \geq d$, we have $\alpha(c'L') + \sum c'' - \min(c'') \geq d$. Because $d' = d - \sum c''$, we derive $\alpha(c'L') + \sum c'' - \min(c'') \geq d' + \sum c''$ which simplifies to $\alpha(c'L') \geq d' + \min(c'')$. □

With Lemma 7.6.12 we can replace the strict reduction $c'L' \geq d'$ by the stronger linear pseudo-Boolean inequality $c'L' \geq d' + \min(c'')$ while preserving equivalence w.r.t. the context S.

Example 7.6.13 Let $cL \geq d$ again be

$$6 \cdot A + 5 \cdot B + 4 \cdot C + 3 \cdot D + 2 \cdot E + F \geq 13$$

and suppose that $S \cup \{B + C \geq 2\}$ is unsatisfiable. A non-redundant extended clause dominated by $cL \geq d$ is $A + E + F \geq 1$. While eliminating B and C we detect unsatisfiability of $S \cup \{B + C \geq 2\}$ and can replace the reduction

$$6 \cdot A + 3 \cdot D + 2 \cdot E + F \geq 4$$

by

$$6 \cdot A + 3 \cdot D + 2 \cdot E + F \geq 8$$

from which we derive the non-redundant extended clause $A + E \geq 1$ dominating $A + E + F \geq 1$. Note that $6 \cdot A + 3 \cdot D + 2 \cdot E + F \geq 8$ dominates $A \geq 1$, which is obtained from the strongest extended clause of the strict reduction obtained when further eliminating $\{D, E, F\}$. So in this case we replace a non-redundant extended clause by a stronger one, but we may also derive new extended clauses that do not dominate other generated extended clauses.

We propose the use of Lemma 7.6.12 in the constrained transformation algorithm and to consider no further strict reductions of the stronger strict reduction $c'L' \geq d' + \min(c'')$, but only add $\mathsf{SCP}(c'L' \geq d' + \min(c''))$[2] to the set of extended clauses. Note that we get the stronger extended clause $\mathsf{SCP}(c'L' \geq d' + \min(c''))$ with minimal computational effort and we have to add $\mathsf{SCP}(c'L' \geq d')$ in any case in order to apply the redundancy criteria presented in Section 7.2.

7.7 CUT GENERATION FOR 0-1 INTEGER PROGRAMMING

Extended cover inequalities of a linear 0-1 inequality, well known in 0-1 integer programming, can be related to the extended clauses generated by our transformation method. In linear programming based 0-1 optimization algorithms (branch-and-bound), valid extended cover inequalities can be added to the current constraint set in order to strengthen the linear programming relaxation of the problem [NW88, NSS94].

We show how to use our transformation method for generating these extended cover inequalities and how to restrict the transformation method in order to produce *strong* extended cover inequalities or extended cover inequalities over a special subset of variables. We test the method on a subset of 0-1 integer programming problems and report computational results.

7.7.1 Extended Clauses and Extended Cover Inequalities

Reformulating 0-1 integer programming problems and applying constraint generation techniques make large scale 0-1 optimization problems solvable in reasonable time [CJP83, JKS85]. The widely used branch-and-bound method for 0-1 integer optimization problems solves the linear programming (LP) relaxation of the problem and then branches on variables with fractional values in the current LP-solution. At each node valid 0-1 inequalities may be generated, so-called strong cuts, that better approximate the integral solutions of

[2] Or better the non-redundant extended clause dominating $\mathsf{SCP}(c'L' \geq d' + \min(c''))$ on the path H_{k+1} of Proposition 7.2.8.

the problem. Therefore, solving the reformulated problems typically requires fewer branch and bound cycles.

The constraint set of a linear 0-1 integer programming problem consists of a set of linear 0-1 inequalities of the form

$$c_1 \cdot X_1 + \cdots + c_n \cdot X_n \leq d \qquad (7.27)$$

which is equivalent to

$$-c_1 \cdot X_1 - \cdots - c_n \cdot X_n \geq -d$$

and then can be brought into pseudo-Boolean normal form.

An important constraint generation technique is presented in [NW88]. It is based mainly on the notion of *minimal covers* or *minimal dependent sets*.

Definition 7.7.1 [NW88] A set $C \subseteq \{1, \ldots, n\}$ is a minimal dependent set of a linear 0-1 inequality (7.27) if

$$\sum_{i \in C \setminus \{j\}} c_i \leq d < \sum_{i \in C} c_i$$

for all $j \in C$.

Proposition 7.7.2 *[NW88] If C is a minimal dependent set for a linear 0-1 inequality $\sum_{i=1}^{n} c_i X_i \leq d$, then*

$$\sum_{i \in E(C)} X_i \leq |C| - 1 \ , \qquad (7.28)$$

where

$$E(C) = C \cup \{k \mid 1 \leq k \leq n \text{ and } c_k \geq c_j \text{ for all } j \in C\}$$

is a valid inequality w.r.t. $\sum_{i=1}^{n} c_i X_i \leq d$. *The set of indices $E(C)$ is called the* extension *of C and we call (7.28)* an extended cover inequality.

Proof. We show that if α is not in $\mathsf{Ext}(\sum_{i \in E(C)} X_i \leq |C| - 1)$ then α is not in $\mathsf{Ext}(\sum_{i=1}^{n} c_i X_i \leq d)$.
Let α be an assignment that is not in $\mathsf{Ext}(\sum_{i \in E(C)} X_i \leq |C| - 1)$. Then $\alpha(\sum_{i \in E(C)} X_i) \geq |C|$. Therefore, $\alpha(\sum_{i=1}^{n} c_i X_i)$ is a sum of at least $|C|$ coefficients c_i with $i \in E(C)$. The sum of $|C|$ coefficients c_i with $i \in E(C)$ is greater than d by construction. Hence, $\alpha(\sum_{i=1}^{n} c_i X_i) > d$, which means $\alpha \notin \mathsf{Ext}(\sum_{i=1}^{n} c_i X_i \leq d)$. □

Note that

$$\sum_{i \in E(C)} X_i \leq |C| - 2$$

is not valid w.r.t. $\sum_{i=1}^{n} c_i X_i) > d$. Let $\alpha(X_i) = 1$ for all $i \in C \setminus \{j\}$ for some $j \in C$ and $\alpha(X_i) = 0$ otherwise. Then $\alpha \notin \mathsf{Ext}(\sum_{i \in E(C)} X_i \leq |C| - 2)$, but $\alpha \in \mathsf{Ext}(\sum_{i=1}^{n} c_i X_i \leq d)$.

Let us write (7.28) in pseudo-Boolean normal form.

$$\sum_{i \in E(C)} \overline{X_i} \geq |E(C) \setminus C| + 1 \tag{7.29}$$

We know that

$$\sum_{i \in E(C)} \overline{X_i} \geq |E(C) \setminus C| + 2$$

is not a valid inequality, because otherwise

$$\sum_{i \in E(C)} X_i \leq |C| - 2$$

would be valid, which is not the case. Hence, (7.29) is the strongest extended clause containing exactly the literals X_i for all $i \in E(C)$. Because each extended cover inequality of $cL \geq d$ corresponds to a strongest extended clause in $\mathsf{SCP}(\mathsf{SRed}(cL \geq d))$, we can use our transformation procedure to systematically construct the set of all strongest *non-redundant* extended cover inequalities.

In a preprocessing phase of linear programming based 0-1 integer programming systems, extended cover inequalities are generated and *added* to the original constraint set. Because of our equivalence result, we can replace the original inequality by its set of equivalent extended clauses. For a branch-and-cut system, where valid constraints may be added at each node, we can generate all extended cover inequalities, but store them separately. At each node we can then select just one, if it violates the current LP-solution.

Example 7.7.3 Let us look again at Example 7.3.3 on page 127.

[a] By writing the extended clauses as "$\leq$-inequalities" we see that the two extended cover inequalities

$$\begin{aligned} X_1 + X_2 + X_3 + X_4 + X_5 &\leq 3 \\ X_1 + X_2 + X_3 &\leq 2 \end{aligned}$$

p0033	# I	# EC	# Nodes	PrePro	Solve	LP-Sol.	Optimum
Original	11	4	9685	0.00	59.89	2520.57	3089
Normal	11	4	1289	0.00	6.47	2819.36	3089
Simple	2	80	124	0.04	1.42	2846.00	3089
Full	0	8710	118	0.90	237.30	2846.00	3089

Figure 7.4 Computational Results for "p0033"

are equivalent to the linear 0-1 inequality

$$79 \cdot X_1 + 53 \cdot X_2 + 53 \cdot X_3 + 45 \cdot X_4 + 45 \cdot X_5 \leq 178 \ .$$

[b] The single extended cover inequality

$$X_1 + X_6 + X_7 + X_9 \leq 1$$

is equivalent to the linear 0-1 inequality

$$774 \cdot X_1 + 76 \cdot X_2 + 22 \cdot X_3 + 42 \cdot X_4 + 21 \cdot X_5 + \\ 760 \cdot X_6 + 818 \cdot X_7 + 62 \cdot X_8 + 785 \cdot X_9 \leq 1500 \ .$$

So for Example 7.7.3[a] and [b] a full transformation is suitable. For Example 7.3.3[c] the number of extended clauses is too large and we need to restrict the number of extended clauses we generate. A trivial, but powerful, approach is to transform only *simple* linear pseudo-Boolean inequalities. The transformation algorithm can easily be changed to just count the number of extended clauses that will be generated, which will be much faster than generating the extended clauses. We can then only transform linear pseudo-Boolean inequalities that generate fewer extended clauses than a given upper bound.

For example, the constraint set of the 0-1 optimization problem p0033 [NW88, page 465] (also available in MIPLIB [BBI92]) consists of 15 inequalities, where 4 of them are already extended clauses. From the remaining 11 inequalities 9 are *simple*, since they are equivalent to only a few extended clauses, and just 2 inequalities produce more than 4000 extended clauses. Thus, a naive approach is to transform the 13 simple inequalities completely in advance and leave the 2 *hard* inequalities unchanged. We ran the commercial mixed integer solver CPLEX 2.1 on the original and the preprocessed problem. The results are reported in Figure 7.4. We denote by "# I" the number of linear pseudo-Boolean inequalities where some coefficients on the left-hand side are greater than 1, and by "# EC" the number of extended clauses. "# Nodes" denotes the number of nodes explored by CPLEX in order to find the optimal solution and to prove optimality. "PrePro" is the time in user cpu seconds on a SPARC-10 used by a PROLOG-implementation of the transformation method in order to preprocess the problem. "Solve" is the time in user cpu seconds CPLEX needed to solve the preprocessed problem. By "LP-Sol." we denote the optimal solution of the linear programming relaxation of the problem and by "Optimum" the integer optimal solution. The rows describe 4 different formulations of "p0033". In "Original" we solve the original problem as found in MIPLIB. In "Normal" we solve the problem after bringing each inequality into pseudo-Boolean normal form.

Bringing an inequality into pseudo-Boolean normal form includes coefficient reduction as presented in [CJP83] and for "p0033" we obtain a speedup of factor 9. This is reflected by the better optimal solution of the linear programming relaxation of the problem and the fewer number of linear programs solved ("Nodes"). Transforming the *simple* inequalities completely generates a larger problem (82 versus 15 constraints), but we obtain an additional speedup of factor 4, which is again due to a better linear programming relaxation. Note that the time consumed by the preprocessing phase is very small w.r.t. the "Solve"-time. A complete transformation of all inequalities shows to be too clumsy. The size of the problem (8710 versus 15 inequalities) slows down the linear programming solver, although even fewer nodes are needed. Using constrained transformation as presented in Section 7.6.3, which means transforming each inequality completely in the context of all remaining inequalities, generates only 4484 extended clauses. So when a complete transformation is required, constrained transformation is advantageous. Complete transformation of a constraint set of a 0-1 integer optimization problem is not always well-suited. Therefore, we need to make our transformation method more flexible.

7.7.2 Restrictions

By restricting the number of generated extended clauses, we can use the transformation method as generator of extended cover inequalities, i.e., strongest extended clauses, in a branch-and-cut system for solving 0-1 integer optimization problems. We show how to focus on extended clauses that are sufficiently strong and restrict the number of extended clauses by focusing on some set of variables.

First, we give a measure of extended clauses describing their *quality* and show how to adopt the transformation method in order to obtain only extended clauses better than or equal to a given quality border. The *weakness* of an extended clause $L \geq d$ is defined by $\mathsf{w}(L \geq d) := |L| - d$ (see Definition 8.1.1). Note that $\mathsf{w}(L \geq d) \leq \mathsf{w}(L' \geq d')$ is a necessary condition for $L \geq d$ dominates $L' \geq d'$, which is obvious from Lemma 3.1.14. We say that $L \geq d$ is *stronger* than $L' \geq d'$ if $\mathsf{w}(L \geq d) \leq \mathsf{w}(L' \geq d')$.

Given a linear pseudo-Boolean inequality, we can easily determine the minimal weakness of all extended clauses the transformation method generates.

Lemma 7.7.4 *Let $cL \geq d$ be a linear pseudo-Boolean inequality in normal form and let β be such that*

$$\sum_{i=1}^{\beta} c_{n-i+1} < d \leq \sum_{i=1}^{\beta+1} c_{n-i+1} \quad , \tag{7.30}$$

then $\beta' = |L| - \beta - 1$ is the smallest weakness of all extended clauses generated by the transformation method.

Proof. We know that

$$\sum_{i=1}^{n-\beta} c_i L_i \geq d - \sum_{i=1}^{\beta} c_{n-i+1}$$

is a valid strict reduction of $cL \geq d$. The right-hand side of its corresponding strongest extended clause is 1 because of (7.30) and therefore its weakness is $n - \beta - 1 = \beta'$. Next, we show that there is no stronger extended clause, i.e., no extended clause with a smaller weakness. Let us recall (7.8), which states that the degrees of the extended clauses along a path are either equal or decrease by one from father node to son node. On the other hand, the number of literals in the extended clauses along a path decrease always by one from father node to son node. So, we conclude that the weakness of the extended clauses along a path are either equal or decrease by one from father node to son node. In order to determine the minimal weakness, it is therefore sufficient to consider only the leaves of the graph, i.e., all extended clauses, where the right-hand side is 1. Hence, we only need to investigate all strict reductions

$$\sum_{i \in M} c_i L_i \geq d - \sum_{i \in N \setminus M} c_i \tag{7.31}$$

with $M \subseteq N$ such that $\sum_{i \in M} c_i < d$ and $c_{\max(N \setminus M)} + \sum_{i \in M} c_i \geq d$. The weakness of these extended clauses is $|M| - 1$ and we search for an M with a minimal cardinality. Obviously, $M' = \{1, \ldots, \beta\}$ has minimal cardinality, since $\sum_{i=1}^{\beta} c_i \geq \sum_{i \in M} c_i$ for all M satisfying (7.31). □

We can easily adapt our transformation method to generate only extended clauses having a weakness less than or equal to a given bound by applying Lemma 7.7.4 to every strict reduction we construct and abandon further strict reductions, if the minimal weakness is greater than the bound. So, by Lemma 7.7.4 we can filter out the generation of extended clauses that do not have a weakness less than or equal to the bound. Note that we avoid unnecessary computation of branches that do not lead to extended clauses that are strong enough.

A very popular and strong class of extended clauses that can be generated this way are extended clauses with weakness 1.

Definition 7.7.5 An extended clause of the form $L \geq |L| - 1$, i.e., an extended clause with weakness 1, is called *clique inequality* or *special ordered set constraint* [NW88].

Clique inequalities have proven to be useful in 0-1 integer programming. With the method presented here in combination with Lemma 7.7.4 we can generate extended clauses having an arbitrary weakness limit and generalize so existing methods that generate only cliques.

Another possibility is to generate only extended clauses from a linear pseudo-Boolean inequality containing a special subset of variables. Suppose that we have a clique inequality for some variables and have no extended clauses containing the remaining variables. So we are interested in generating extended clauses, where this set of remaining variables V occurs. Another interesting set of variables is the one with large coefficients in the objective function, since they are mainly responsible for the resulting objective function value.

We obtain valid extended clauses containing a special set V of variables, if we transform a strict reduction of the linear pseudo-Boolean inequality, where we have eliminated in advance all literals containing the variables that are not in V. Another reason to transform a strict reduction and not the whole linear pseudo-Boolean inequality is given, if the linear pseudo-Boolean inequality contains some relatively large and some relatively small coefficients. The generated strongest extended clauses typically contain the literals with the largest coefficients and eliminating literals with small coefficients in advance focuses on

Name	# I	# EC	# Nodes	PrePro	Solve	LP-Sol.	Opt.
p0033-Original	11	4	9685	0.00	59.89	2520.57	3089
p0033-Normal	11	4	1289	0.00	6.47	2819.36	3089
p0033-Simple	2	80	124	0.04	1.42	2846.00	3089
p0040-Original	23	20	96	0.00	0.55	61796.54	62027
p0040-Normal	23	20	59	0.00	0.32	61829.08	62027
p0040-Simple	0	24	0	0.03	0.05	62027.00	62027
p0201-Original	33	100	5238	0.00	167.95	6875.00	7615
p0201-Normal	33	100	2775	0.00	90.42	7125.00	7615
p0201-Simple f	30	106	2273	0.33	84.83	7125.00	7615
p0201-wl(9) f	30	112	1258	0.43	47.47	7125.00	7615
p0548-Original	89[1]	63	&	0.00	&	315.25	8691
p0548-Normal	89	63	&	0.02	&	4568.75	8691
p0548-Simple f	52	508	&	0.54	&	6617.47	8691
p0548-wl(1) cr f	52	672	&	8.10	&	7306.68	8691
p2756-Original	378[2]	352	&	0.00	&	2688.75	3124
p2756-Normal	378	352	&	2.35	&	2701.14	3124
p2756-wl(1) cr f	343	2096	&	256.50	&	2701.75	3124
lseu-Original	11	17	55526	0.00	791.60	834.68	1120
lseu-Normal	11	17	28551	0.00	465.65	944.75	1120
lseu-Simple	7	22	18032	0.00	275.02	985.54	1120

[1] From 176 inequalities are 10 tautologies and 14 become tautologous after fixing
[2] From 755 inequalities are 6 tautologies and 19 become tautologous after fixing
& out of memory

Figure 7.5 Some Computational Results

these strong extended clauses. Note that we can obtain strong extended clauses also by restricting the weakness as showed at the beginning of this section. By eliminating literals in advance, we reduce the size of the search tree of the transformation method, but in exchange only approximate the strongest extended clauses. The approach can be used if the linear pseudo-Boolean inequalities are too large to be transformed completely.

7.7.3 Computational Results

We have tested the preprocessing approach on some pure 0-1 integer programming problems found in MIPLIB [BBI92]. For the format of the figures see Section 7.7.1, page 153. Additionally, we denote by "f" whether variables have been fixed using the `fixing` procedure presented in Section 7.5, and with "wl(w)"

that for the non-simple inequalities the generation of extended clauses has been limited to the maximal weakness w.

We obtain encouraging results on the "pxxxx" problem set [CJP83], since these are sparse problems and considering each inequality on its own is promising under this premise. On dense problems we cannot expect similar improvements. We denote by "cr" whether coefficient reduction, as presented in Section 7.6.1, has been successfully applied to problem. Coefficient reduction is tried on the literal with the largest coefficient in each linear pseudo-Boolean inequality that cannot be simply transformed. The set of extended clauses that is used for the coefficient reduction is the set of extended clauses generated by the transformation of all simple linear pseudo-Boolean inequalities. Coefficient reduction was not very successful in the preprocessing phase. Coefficient reduction would be more successful when applied on the linear pseudo-Boolean inequalities constructed while exploring the branch-and-bound tree, since then some literals are fixed and so better bounds may be obtained for the simpler linear pseudo-Boolean inequalities.

The presented techniques can be used in a branch-and-bound method when solving 0-1 integer programming problems. On spare problems the techniques are useful even in a preprocessing phase. Significant speedups can be obtained when the optimal solution of the linear programming relaxation of the partially transformed problem is better than in the original problem. In the context of 0-1 integer optimization we generalize the generation of clique inequalities (extended clauses with weakness 1) to the generation of extended clauses with an arbitrary weakness limit. The transformation method can be used as a cut generator in a branch-and-bound algorithm. When searching an extended clause violating the fractional solution of the current linear programming relaxation, we can extract just the needed extended clause. With the given redundancy criteria we assure that we select the best possible extended clause. For example, assume that the variables A, B have the fractional value 0.5 in the current LP-solution, and that a linear pseudo-Boolean inequality $cL \geq d$ of the constraint set dominates the non-redundant extended clause $A + B + C \geq 2$, i.e., $A + B + C \geq 2$ is in $\mathsf{SCP}(\mathsf{SRed}(cL \geq d))$. When searching some polyhedral cut that violates the current fractional solution, we may just compute the extended clause $A + B \geq 1$, which is also in $\mathsf{SCP}(\mathsf{SRed}(cL \geq d))$, and add it to the problem. Because of the redundancy criteria, we choose with our approach $A + B + C \geq 2$ and so forbid further fractional solutions, among them, for example, $B = C = 0.5$.

Note that the main feature of the transformation method is that on the obtained set of extended clauses the deductive system *Generalized Resolution* [Hoo92a] can be applied. Thus, information coming from several linear pseudo-Boolean inequalities are used and may lead to stronger extended clauses. In Chapter 8 we investigate how to strengthen extended clauses by using a restricted version of *Generalized Resolution.* The output of the transformed and strengthened constraint set can be stored separately and extended clauses serving as cuts then can be selected when needed by the branch-and-cut algorithm.

7.8 IMPLEMENTING RESOLUTION

Hooker [Hoo92a] defines a *resolvent* of two extended clauses as the classical resolvent of two classical clauses, each of them dominated by one of the extended clauses. We enhance the notion of resolvents between extended clauses and show how to efficiently implement the inference rule with the help of our transformation method.

Example 7.8.1 Let

$$A + B + \overline{D} + E + \overline{F} + \overline{G} \geq 4 \tag{7.32}$$

$$\overline{A} + \overline{C} + \overline{D} + \overline{E} + F + G \geq 3 \tag{7.33}$$

be two extended clauses. The classical clause $A + B + \overline{D} \geq 1$ is dominated by (7.32) and $\overline{A} + \overline{C} + \overline{D} + \overline{E} \geq 1$ is dominated by (7.33). From these two classical clauses we obtain the classical resolvent $B + \overline{C} + \overline{D} + \overline{E} \geq 1$ by resolving on A.

As shown in Section 3.3, the resolvent of two classical clauses can be obtained by linear combination of the two classical clauses and the valid bounds $L_i \geq 0$ and $-L_i \geq -1$ for all literals L_i, and integer rounding. Thus, resolvents are also cutting plane inequalities. It is easy to see that resolvents of extended clauses can also be obtained by linear combination and rounding, since classical clauses dominated by extended clauses are strict reductions that are obtained by adding the valid bound $-L_i \geq -1$ for all literals L_i that are eliminated. In the following we show that each resolvent of two extended clauses is dominated by an extended clause obtained by transforming the sum of these two extended clauses. The *sum* of two linear pseudo-Boolean inequalities $cL \geq d$ and $c'L' \geq d'$ is the linear pseudo-Boolean inequality $cL + c'L' \geq d + d'$. As usual, we assume that the sum of two linear pseudo-Boolean inequalities is brought into pseudo-Boolean normal form.

Proposition 7.8.2 *Each resolvent of two extended clauses $L \geq d$ and $L' \geq d'$ is dominated by an extended clause in* $\mathsf{SCP}(\mathsf{SRed}(cL + c'L' \geq d + d'))$.

Proof. We construct two classical clauses that are dominated by $L \geq d$ resp. $L' \geq d'$ and have a resolvent. We then show that this resolvent is dominated by an extended clause in $\mathsf{SCP}(\mathsf{SRed}(cL + c'L' \geq d + d'))$.

First, we extract the literals that occur positively in one extended clause and negatively in the other. Let $L_R + K = L$ and $L'_R + \overline{K} = L'$ with $K = \{L_i \mid L_i \in L \text{ and } \overline{L_i} \in L'\}$ and $\overline{K} = \{\overline{L_i} \mid L_i \in K\}$. Next, we construct two classical clauses that are dominated by $L \geq d$ resp. $L' \geq d'$ and have a resolvent. The general representation[3] of such a classical clause that is dominated by $L \geq d$ is $L_C + K_C \geq 1$ with $|K_C| = 1$ and $|L_R \setminus L_C| + |K \setminus K_C| \leq d - 1$, where $L_C \subseteq L_R$ and $K_C \subseteq K$. Similarly, the general representation of such a classical clause that is dominated by $L' \geq d'$ is $L'_C + K'_C \geq 1$ with $|K'_C| = 1$ and $|L'_R \setminus L'_C| + |\overline{K} \setminus K'_C| \leq d' - 1$, where $L'_C \subseteq L'_R$ and $K'_C \subseteq \overline{K}$. Furthermore, $K_C = \overline{K'_C}$ must hold. Now, we extract the literals that occur in both L_C and L'_C. Let $P \cup Q = L_C$ and $P \cup Q' = L'_C$ with $L_C \cap L'_C = P$ and $Q \cap Q' = \emptyset$. So, the general representation of a classical resolvent of $L \geq d$ and $L' \geq d'$ is

$$P \cup Q \cup Q' \geq 1 \ .$$

Since $|L_R \setminus L_C| + |K \setminus K_C| \leq d - 1$ and $|L'_R \setminus L'_C| + |\overline{K} \setminus K'_C| \leq d' - 1$ we have

$$d + d' \geq |L_R \setminus L_C| + |L'_R \setminus L'_C| + 2 + 2 \cdot (|K| - 1) \ ,$$

which simplifies to

$$d + d' \geq |L_R \setminus L_C| + |L'_R \setminus L'_C| + 2 \cdot |K| \ . \tag{7.34}$$

Let us investigate the sum of $L \geq d$ and $L' \geq d'$. We know that $L + L' \geq d + d'$ is equivalent to

$$L_R + L'_R + |K| \geq d + d' \ .$$

A strict reduction of $L + L' \geq d + d'$ is

$$L_C + L'_C \geq d + d' - |K| - |L_R \setminus L_C| - |L'_R \setminus L'_C| \ ,$$

obtained by eliminating the literals in $(L_R \setminus L_C) \cup (L'_R \setminus L'_C)$, which dominates

$$L_C + L'_C \geq |K| \ ,$$

[3] Note that we consider only classical clauses that contain only literals occurring in the extended clause dominating it, because otherwise their resolvent is dominated by either one of the fathers, or by the resolvent of the valid classical clauses that contain only literals occurring in the extended clause dominating it.

because of (7.34). Note that $|K| \geq 1$. We rewrite L_C and L'_C and obtain

$$2 \cdot P + Q + Q' \geq |K| \ .$$

We apply the cutting plane operation of Section 7.1 and obtain

$$P + Q + Q' \geq \left\lceil \frac{|K|}{2} \right\rceil \geq \left\lceil \frac{1}{2} \right\rceil = 1 \ .$$

Hence, $P+Q+Q' \geq 1$ is dominated by the extended clause $P+Q+Q' \geq \lceil |K|/2 \rceil$. Because each extended clause obtained by linear combination and rounding is dominated by a strongest extended clause, the proposition is established. □

By Proposition 7.8.2 we can replace the original resolution rule of *Generalized Resolution* by a rule generating the equivalent set of extended clauses of the linear pseudo-Boolean inequality $L+L' \geq d+d'$ while preserving completeness.

Example 7.8.3 [Hoo92a] We again consider Example 7.8.1. Let

$$\begin{array}{c} A + B + \overline{D} + E + \overline{F} + \overline{G} \geq 4 \\ \overline{A} + \overline{C} + \overline{D} + \overline{E} + F + G \geq 3 \end{array}$$

be two extended clauses. With the enhanced resolution rule of Hooker [Hoo92a] we may generate 12 different resolvents, where

$$B + \overline{C} + \overline{D} + \overline{E} + F + \overline{G} \geq 2$$

is one of them. The sum of the two extended clauses brought in normal form is $2 \cdot \overline{D} + B + \overline{C} \geq 3$. After the transformation we obtain the two extended clauses

$$\overline{D} + \overline{C} + B \geq 2 \qquad \overline{D} \geq 1 \ ,$$

where each of the 12 resolvents is dominated by one of them.

Using the presented transformation method we generate fewer and stronger extended clauses than with the original resolution method. Note that there are no further non-redundant resolution steps, since resolution between the extended clauses in $\mathsf{SCP}(\mathsf{SRed}(L + L' \geq d + d'))$ is not possible, because no literal occurs negatively in one and positively in another of them. Additionally, there is no literal occurring positively in $L \geq d$ or $L' \geq d'$ and negatively in an extended clause in $\mathsf{SCP}(\mathsf{SRed}(L + L' \geq d + d'))$. Thus, the only possible resolution steps are the ones between $L \geq d$ and $L' \geq d'$. But these are all dominated by an extended clause in $\mathsf{SCP}(\mathsf{SRed}(L + L' \geq d + d'))$, because of Proposition 7.8.2. Note that we generate not only resolvents but also *diagonal sums* [Hoo92a], generated by the other rule of *Generalized Resolution.* The sum of $A+B \geq 1$ and $C+D \geq 1$ is $A+B+C+D \geq 2$, which is already an extended clause. Since there is no other extended clause dominating $A + B + C + D \geq 2$ that is derivable from $A + B \geq 1$ and $C + D \geq 1$, the strongest extended clause $A + B + C + D \geq 2$ is non-redundant.

8

SIMPLIFICATION

Simplifying a constraint w.r.t. a set of given constraints is a powerful constraint solving technique, particularly in the context of constraint logic programming. In this chapter we consider only simplification techniques for a set of extended clauses. Simplification techniques for linear pseudo-Boolean inequalities were introduced in Section 7.6 (as they fit better into a chapter dealing mainly with linear pseudo-Boolean inequalities).

The basic idea of the simplification rules in this chapter is that we take an inference rule generating a valid extended clause $L \geq d$ from a set of extended clauses S and then restrict its application to the case that $L \geq d$ strictly dominates at least one other extended clause in S. Thus, an inference rule $S \to L \geq d$ is restricted to the case where

$$\exists L' \geq d' \in S : L \geq d \text{ strictly dominates } L' \geq d'$$

holds. We say that $L \geq d$ is a *simplifying* extended clause. Given a simplifying extended clause $L \geq d$, we replace the dominated extended clause $L' \geq d'$ by $L \geq d$ without changing the set of solutions. The idea is that we explicitly generate an extended clause such that an existing extended clause in S becomes *redundant* and thus we *simplify* the set of constraints.

Assumption 8.0.1 Throughout this chapter we assume that all sets of extended clauses are in *normal form*, i.e., in a set of extended clauses there is no extended clause dominating another extended clause in that set.

We define a complete order for extended clauses in Section 8.1 and show that the simplification rules generate smaller extended clauses w.r.t. that order. In Section 8.2 we restrict the diagonal sum rule to be simplifying and show in detail how to efficiently obtain a simplifying extended clause. Furthermore, we apply the simplification to combinatorial clausal satisfiability problems, which are solved with traditional 0-1 integer programming techniques, since diagonal sums are typically also strong cuts w.r.t. $[0,1]^n$. The simplification idea is applied to the resolution rule in Section 8.3.

We show how to detect equations between two Boolean variables in Section 8.4. Since an equation between two Boolean variables is not obviously representable given only extended clauses as basic formulas, equality detection by a special treatment helps to simplify the constraint store.

8.1 AN ORDER FOR EXTENDED CLAUSES

We define a complete ordering for extended clauses that is an approximation of the domination relation.

Definition 8.1.1 We denote by $\mathsf{w}(L \geq d) := |L| - d$ the *weakness* of an extended clause $L \geq d$.

By rewriting an extended clause $L \geq d$ as a $\leq$-inequality we get $\overline{L} \leq \mathsf{w}(L \geq d)$ with $\overline{L} = \{\overline{L_i} \mid L_i \in L\}$. Note that $\mathsf{w}(L \geq d) + 1$ is the number of literals of all shortest classical clauses dominated by C.

Lemma 8.1.2 *If an extended clause $L \geq d$ dominates an extended clause $L' \geq d'$, then $\mathsf{w}(L \geq d) \leq \mathsf{w}(L' \geq d')$.*

Proof. If $L \geq d$ dominates $L' \geq d'$, then we have $|L \setminus L'| \leq d - d'$ by Lemma 3.1.14. Since $|L| - |L'| \leq |L \setminus L'|$, we know that $|L| - |L'| \leq d - d'$, which is equivalent to $|L| - d \leq |L'| - d'$ and therefore $\mathsf{w}(L \geq d) \leq \mathsf{w}(L' \geq d')$. □

We define the following irreflexive and transitive ordering over $\mathcal{L}$. Let ϕ be the function from $\mathcal{L}$ to $\mathbb{N}$ with $\phi(X_j) = 2j$ and $\phi(\overline{X_j}) = 2j + 1$. Then ϕ maps each literal to a unique natural number. We say a literal L_i is smaller than a literal L'_i (written $L_i \lhd L'_i$) if and only if $\phi(L_i) < \phi(L'_i)$. Let us view a set of literals L as an ordered n-tuple $(L_1, \ldots, L_n)$ with $L_i \lhd L_j$ if $i < j$. We then have the total ordering $\ll$ for sets of literals, where $\ll$ is the lexicographic ordering over $\lhd$. We define the following total ordering for extended clauses.

Definition 8.1.3 We say that an extended clause $L \geq d$ is *smaller* than an extended clause $L' \geq d'$ (written $L \geq d \prec L' \geq d'$) if

- $\mathsf{w}(L \geq d) < \mathsf{w}(L' \geq d')$ or
- $\mathsf{w}(L \geq d) = \mathsf{w}(L' \geq d')$ and $d > d'$ or
- $\mathsf{w}(L \geq d) = \mathsf{w}(L' \geq d')$ and $d = d'$ and $L \ll L'$.

We use $\preceq, \succ$ and $\succeq$ with their obvious meaning. For classical clauses the ordering reduces to the weakness, the length of a classical clause, i.e., the number of its literals, and the lexicographic ordering $\ll$. This is the usual ordering for classical clauses.

Proposition 8.1.4 *If $L \geq d$ dominates $L' \geq d'$, then $L \geq d \preceq L' \geq d'$.*

Proof. Suppose that $L \geq d$ dominates $L' \geq d'$, then from Lemma 8.1.2 we know that $\mathsf{w}(L \geq d) \leq \mathsf{w}(L' \geq d')$ and from Lemma 3.1.14 we have $d \geq d'$. Suppose that $L' \geq d' \prec L \geq d$, then $\mathsf{w}(L \geq d) = \mathsf{w}(L' \geq d')$, $d = d'$ and therefore $|L| = |L'|$. Because of Lemma 3.1.14, we have $|L \setminus L'| = d - d' = 0$, and thus $L = L'$. But then $L' \geq d'$ equals $L \geq d$, which contradicts $L' \geq d' \prec L \geq d$. Therefore, $L \geq d \preceq L' \geq d'$. □

Note that the ordering is not well founded given an infinite number of Boolean variables. For example,

$$\sum_{1 \leq i \leq n} X_i \geq n \quad \text{is dominated by} \quad \sum_{1 \leq i \leq n+1} X_i \geq n+1$$

for all $n > 0$, but the latter extended clause is smaller than the former. We assume that the number of variables, and thus the number of literals, is finite, which is reasonable for all practical applications, and then the ordering $\prec$ is well founded.

8.2 SIMPLIFYING DIAGONAL SUMS

First, we restate the diagonal sum rule of *Generalized Resolution* [Hoo92a] for extended clauses and then restrict its application to generate only simplifying diagonal sums.

Definition 8.2.1 [Hoo92a] An extended clause $DS := \sum_{1 \leq i \leq n} L_i \geq d + 1$ is a *diagonal sum* of a set S of extended clauses if

[a] for all $1 \leq j \leq n$ the extended clause $DS_j := \sum_{1 \leq i \neq j \leq n} L_i \geq d$ is dominated by an extended clause in S,

[b] there is no extended clause in S that dominates DS, and

[c] $\mathsf{Var}(L_i) \neq \mathsf{Var}(L_j)$ for all $i \neq j$.

We say S *supports* DS. A set $S' \subseteq S$ supporting DS is called *minimal support* of DS if S' supports DS but no proper subset of S'. We denote by $DSF := \{DS_j \mid 1 \leq j \leq n\}$ the set of *fathers* of DS.

If a set of extended clauses S supports a diagonal sum DS, then there exists a minimal supporting set $S' \subseteq S$.

Proposition 8.2.2 *A diagonal sum DS of a set of extended clauses S is a valid extended clause with respect to S.*

Proof. Summing up all the DS_j yields the linear pseudo-Boolean inequality $(n-1) \cdot L \geq n \cdot d$. Dividing this inequality by $n-1$ yields $L \geq n \cdot d/(n-1)$. Because each L_i is either 0 or 1, the sum of the L_i is integral. Therefore, it is valid to round up the right side to the next integral value. Since $n/(n-1) > 1$, we have $\lceil n \cdot d/(n-1) \rceil \geq d + 1$. □

Note that a diagonal sum is a *rank 1 Chvátal-Gomory cutting plane* [Gom58, NW88], obtained from a positive linear combination of valid inequalities and integer rounding.

Definition 8.2.3 Let S be a set of extended clauses supporting a diagonal sum DS. If there is an extended clause $C \in S$ that is dominated by DS, then we call DS a *simplifying diagonal sum* with respect to S.

Given a simplifying diagonal sum DS of S, we can add DS to S and delete dominated extended clauses in S yielding a *simpler* set S'. Note that $\mathsf{Ext}(S') = \mathsf{Ext}(S)$ and $|S'| \leq |S|$, since at least one extended clause $C \in S$ is dominated by DS. Since DS is smaller than C w.r.t. the ordering $\prec$ of Definition 8.1.3, and thus we replace a larger element by a smaller one, we have another justification of the simplification argument. Because $\prec$ is well founded, we know that the simplification algorithm terminates, and that there is a fully simplified form.

The simplification argument is also valid w.r.t. $[0,1]^n$. If we compare the polytope of the linear programming relaxation of the set of fathers of the simplifying diagonal sum and the linear programming relaxation of the simplifying diagonal sum, then we see that the latter is included in the former, i.e., the latter better approximates the convex hull of the 0-1 solutions points. To see this assign to each $L_i \in DS$ the value $d/(n-1)$, then each DS_j is satisfied but not DS. We say we have simplified the problem w.r.t. $[0,1]^n$. In many cases we even cut off solutions of the linear programming relaxation of S and then S' is a better approximation[1] of $\mathsf{Ext}(S)$ (cf. [HF90]).

Example 8.2.4 Let S be $\{X_1+X_2 \geq 1, X_1+X_3+X_4 \geq 1, X_2+X_3+X_4 \geq 1\}$. By adding the diagonal sum $X_1+X_2+X_3+X_4 \geq 2$ and deleting the dominated extended clauses we obtain a simplified set $S' = \{X_1+X_2+X_3+X_4 \geq 2, X_1+X_2 \geq 1\}$. Then $X_1 = X_2 = 0.5$ and $X_3 = X_4 = 0.3$ is a solution of the linear programming relaxation of S but not of the linear programming relaxation of S'.

Instead of first generating diagonal sums and afterwards checking whether there is one dominating an extended clause, which is rather inefficient, we show how to search for a simplifying diagonal sum dominating a specific extended clause of the current set of extended clauses. In other words, we pick an extended clause of the set of extended clauses and try to construct a valid diagonal sum dominating it.

8.2.1 Properties of Simplifying Diagonal Sums

We show some useful properties of simplifying diagonal sums. First, we want to restrict the search for fathers of a simplifying diagonal sum.

Proposition 8.2.5 *Let S be a set of extended clauses supporting the diagonal sum $L \geq d+1$ and let $S' \subseteq S$ be a minimal support of $L \geq d+1$. Then for all $L' \geq d' \in S'$ we have $\mathsf{w}(L' \geq d') \leq \mathsf{w}(L \geq d+1)$ and $d' \geq d$.*

[1] Note that in general we do not cut off solutions of the linear programming relaxation. For example, suppose that the linear programming relaxation is infeasible (let S contain $A+B+C \geq 2$ and $\overline{A}+\overline{B}+\overline{C} \geq 2$), then we cannot cut off anything.

Proof. If S' is a minimal support, then each $L' \geq d' \in S'$ dominates at least one $DS_i := L \setminus \{L_i\} \geq d$. Because of Proposition 8.1.4, we know that $L' \geq d' \preceq DS_i$, and therefore $\mathsf{w}(L' \geq d') \leq \mathsf{w}(DS_i) = \mathsf{w}(DS)$ and $d' \geq d$. □

Given an extended clause $L' \geq d' \in S$, we are searching for a simplifying diagonal sum $L \geq d+1$ supported by S that dominates $L' \geq d'$. We restrict ourselves to simplifying diagonal sums $L \geq d+1$ with $d+1 > d'$, i.e., we want to replace $L' \geq d'$ by an extended clause with a greater right-hand side.

Next, we show that if there is a diagonal sum $L \geq d+1$ supported by a set of extended clauses S such that $L \geq d+1$ dominates the extended clause $L' \geq d' \in S$ with $d+1 > d'$, then it is sufficient to search for a specific simplifying diagonal sum dominating $L' \geq d'$.

Proposition 8.2.6 *Let S be a set of extended clauses supporting a diagonal sum $L \geq d+1$ that dominates an extended clause $L' \geq d' \in S$ with $d+1 > d'$. Then S supports a diagonal sum of the form $\{L_?\} \cup L' \geq d'+1$, where $L_? \in L \setminus L'$.*

Proof. Since $L \geq d+1$ dominates $L' \geq d'$, we have $|L \setminus L'| \leq d+1-d'$ by Lemma 3.1.14. We split L into two disjoint sets M and N, such that $M \subseteq L'$, $N \cap L' = \emptyset$ and $M \cup N = L$. Note that $M = L \setminus L'$. Since

$$|L \setminus L'| = |(M \cup N) \setminus L'| = |M \setminus L'| + |N \setminus L'| = |N| \leq d+1-d' \ ,$$

we have

$$|M \setminus L'| = 0 \leq d+1-|N|-d' \tag{8.1}$$

and thus $L \geq d+1$ dominates its reduction $M \geq d+1-|N|$, which dominates $L' \geq d'$, because $M \subseteq L'$ and $d+1-|N| \geq d'$.

Next, we show that $N \neq \emptyset$. Suppose the converse, then $M = L \subseteq L'$. Since

$$|(L \setminus \{L_j\}) \setminus L'| = 0 \leq d-d'$$

because $d \geq d'$, we know that $L' \geq d'$ is dominated by a father $L \setminus \{L_j\} \geq d$ of $L \geq d+1$ for any $L_j \in L$. Hence, $L' \geq d'$ is dominated by an extended clause in S, which contradicts the normal form condition and therefore $N \neq \emptyset$.

Let $L_?$ be an arbitrary literal in N. It remains to show that each extended clause

$$\{L_?\} \cup (L' \setminus \{L_i\}) \geq d' \tag{8.2}$$

for all $L_i \in L'$ is dominated by an extended clause in S, since then the diagonal sum rule generates $\{L_?\} \cup L' \geq d'+1$. Because $L \geq d+1$ is supported by S, we know that each extended clause $L \setminus \{L_j\} \geq d$ for all $L_j \in L$ is dominated by an extended clause in S. We show that the extended clauses (8.2) for all $L_i \in L'$ are dominated by a father $L \setminus \{L_j\} \geq d$ of $L \geq d+1$ for some $L_j \in L$. Thus, we have to show for each $L_i \in L'$ that there exists an $L_j \in L$ such that

$$|(L \setminus \{L_j\}) \setminus (L_? \cup (L' \setminus \{L_i\}))| \leq d - d' \ ,$$

which is equivalent to

$$|((M \cup N) \setminus \{L_j\}) \setminus (L_? \cup (L' \setminus \{L_i\}))| \leq d - d' \ .$$

We choose L_j such that $L_? \neq L_j$. Hence, we have to show

$$|N| - 1 + |(M \setminus \{L_j\}) \setminus (L' \setminus \{L_i\})| \leq d - d'$$

We choose $L_j = L_i$ and since $M \subseteq L'$, we obtain

$$|N| - 1 + 0 \leq d - d' \ ,$$

which is equivalent to

$$0 \leq d + 1 - |N| - d' \ .$$

But $0 \leq d + 1 - |N| - d'$ is (8.1). □

8.2.2 Finding Simplifying Diagonal Sums

We discuss now how to construct a simplifying diagonal sum DS of S dominating a given extended clause $L \geq d \in S$. Because of Proposition 8.2.6, we know that it is sufficient to search for simplifying diagonal sums of the form $\{L_?\} \cup L \geq d+1$ for some literal $L_?$. In the following we abbreviate $\{L_?\} \cup L$ by $L_? + L$ and thus

$$DS \equiv L_? + L \geq d + 1 \ . \tag{8.3}$$

The goal is to find a literal $L_?$ such that each extended clause in

$$DSG := \{L_? + \sum_{1 \leq i \neq j \leq n} L_i \geq d \mid \text{ for all } 1 \leq j \leq n\} \tag{8.4}$$

is dominated by an extended clause in S, i.e., we search for a supporting set $S' \subseteq S$ of DS for some $L_?$. Note that we can identify each extended clause in DSG by the missing literal L_j. We show that each extended clause in S that dominates an extended clause in DSG has a special form. Hence, we can restrict the number of extended clauses that need to be investigated.

Proposition 8.2.7 *If $L' \geq d' \in S$ dominates an extended clause in DSG, then $L' \geq d'$ is of the form*

$$(L \setminus M) \cup N \geq d + |N| - 1 \tag{8.5}$$

with $M \subset L$, $|M| \geq 1$, $N \cap L = \emptyset$, and $L_? \in N$.

Proof. We know that $L' \geq d'$ must dominate an extended clause in DSG of the form $L_? + (L \setminus \{L_j\}) \geq d$ with $L_j \in L$. Note that $L_? \in L'$ because otherwise $L' \geq d'$ dominates $L \setminus \{L_j\} \geq d$ and therefore the given extended clause $L \geq d$, which contradicts the normal form condition. We split L' into $(L \setminus M) \cup N$ such that N contains the literals in L' that are not in L and M contains the literals of L that are missing in L'. Note that then $M \subseteq L$, and $L \cap N = \emptyset$ and $L_? \in N$.

First, suppose that $M = \emptyset$. Therefore, we have $L' = L \cup N$. Since $L' \geq d'$ dominates $L_? + (L \setminus \{L_j\}) \geq d$ for some $L_j \in L$, we have

$$|(L \cup N) \setminus (L_? + (L \setminus \{L_j\}))| \leq d' - d \ .$$

Because

$$|(L \cup N) \setminus (L_? + (L \setminus \{L_j\}))| = |N| = |(L \cup N) \setminus L| \ ,$$

we have $|(L \cup N) \setminus L| \leq d' - d$. Therefore, $L \cup N \geq d'$ dominates $L \geq d$, which contradicts the normal form condition. Hence, we know that $|M| \geq 1$.

Since $L' \geq d'$ dominates $L_? + (L \setminus \{L_j\}) \geq d$ for some $L_j \in L$, we have

$$|((L \setminus M) \cup N) \setminus (L_? + (L \setminus \{L_j\}))| \leq d' - d \ .$$

Because $L_? \in N$ and N is disjoint with L, we have $|\{L_j\} \setminus M| \leq d' - d - (|N| - 1)$ and therefore

$$d' \geq d + |N| - 1 + |\{L_j\} \setminus M| \ . \tag{8.6}$$

Suppose that $d' > d + |N| - 1$, then $L' \geq d'$ dominates $L_? + (L \setminus M) \geq d + 1$, since

$$\begin{aligned} |((L \setminus M) \cup N) \setminus (L_? + (L \setminus M))| &= |N \setminus \{L_?\}| \\ &= |N| - 1 \\ &= d + |N| - (d+1) \leq d' - (d+1) \ . \end{aligned}$$

Therefore, $L' \geq d'$ dominates DS and $L \geq d$, which contradicts the normal form condition. We conclude that $L_j \in M$, because of (8.6), and $d' = d + |N| - 1$. □

Note that while searching for extended clauses $L' \geq d'$ dominating an element of DSG, it is sufficient to consider only extended clauses in S, for which condition (8.5) holds. If we have found such an extended clause of the form $(L \setminus M) \cup N \geq d + |N| - 1$, then it dominates all extended clauses $\{L_? + L \setminus \{L_j\} \geq d \mid L_? \in N \text{ and } L_j \in M\}$. We define now the set of possible literals (candidates) for $L_?$ provided by the investigation of each father DS_j.

$$\mathsf{Cand}_{L_j} := \{L_? \mid (L \setminus M) \cup N \geq d + |N| - 1 \in S, L_? \in N \text{ and } L_j \in M\} \quad (8.7)$$

The set of literals Cand_{L_j} is the complete set of literals that can be substituted for $L_?$ such that the father clause DS_j is dominated by an extended clause in S. We see immediately that DS is valid for a specific $L_?$, if substituting it yields a dominated father for all DS_j, i.e., if

$$L_? \in \bigcap_{j=1}^{n} \mathsf{Cand}_{L_j} . \quad (8.8)$$

Example 8.2.8 From the set of extended clauses

$$S := \{X_1 + X_2 \geq 1, X_2 + X_3 \geq 1, X_2 + X_4 \geq 1, X_1 + X_3 + X_5 \geq 2\}$$

we select $X_1 + X_2 \geq 1$ and show how to derive a simplifying diagonal sum for it. We start calculating Cand_{X_1}, i.e., we search for literals $L_?$ such that $L_? + X_2 \geq 1$ is dominated by an extended clause in S. From $X_2 + X_3 \geq 1$ we see that X_3 is a valid choice for $L_?$. Similarly, we see that X_4 is a valid choice for $L_?$ from the extended clause $X_2 + X_4 \geq 1$. The other extended clauses do not satisfy (8.5) and thus we know that $\mathsf{Cand}_{X_1} = \{X_3, X_4\}$. When calculating Cand_{X_2} we see that condition (8.5) only applies for $X_1 + X_3 + X_5 \geq 2$ and we know that X_3 and X_5 are valid choices for the literals $L_?$ such that $L_? + X_1 \geq 1$ is dominated by an extended clause in S. Therefore, $\mathsf{Cand}_{X_2} = \{X_3, X_5\}$ and with (8.8) we have $L_? = X_3$. We then know that $X_1 + X_2 \geq 1$,$X_1 + X_3 \geq 1$, and $X_2 + X_3 \geq 1$ are all dominated by extended clauses in S. Hence, we derive with the diagonal sum rule the extended clause $X_1 + X_2 + X_3 \geq 2$. We add the diagonal sum $X_1 + X_2 + X_3 \geq 2$ of these 3 extended clauses to S and by construction we know that $X_1 + X_2 \geq 1$ is dominated by it. Furthermore, $X_2 + X_3 \geq 1$ is also dominated and both are eliminated. We obtain the simpler set of extended clauses

$$\{X_1 + X_2 + X_3 \geq 2, X_2 + X_4 \geq 1, X_1 + X_3 + X_5 \geq 2\} .$$

8.2.3 Improved Diagonal Sum Rule

In Definition 8.2.1 we require that all DS_j are dominated by an extended clause in S. We show that only a subset of these extended clauses is needed.

Proposition 8.2.9 *An extended clause $DS := L \geq d+1$ is valid with respect to a set of extended clauses S if $n-d+1$ extended clauses of DSF are dominated by extended clauses in S.*

Proof. Suppose that $n-d+1$ extended clauses of DSF are dominated by extended clauses in S. W.l.o.g. suppose that the set of these dominated extended clauses is $DSF' = \{DS_j \mid 1 \leq j \leq n-d+1\} \subseteq DSF$. Summing up the extended clauses in DSF' yields

$$(n-d+1) \cdot \sum_{1 \leq i \leq d-1} L_i + (n-d) \cdot \sum_{d \leq i \leq n} L_i \geq (n-d+1) \cdot d \ .$$

We add the valid bounds $-L_i \geq -1$ for all $1 \leq i \leq d-1$ and obtain

$$(n-d) \cdot \sum_{1 \leq i \leq n} L_i \geq (n-d+1) \cdot d - (d-1) = (n-d) \cdot d + 1 \ .$$

Dividing this inequality by $(n-d)$ yields $L \geq d + 1/(n-d)$. We reason as in Proposition 8.2.2 and, since $1/(n-d) > 0$, we round up the right-hand side to $d+1$ and obtain DS. □

Note that the computed diagonal sum is still a Chvátal-Gomory rank 1 cut. Using Proposition 8.2.9, we can relax condition (8.8). We know that DS is valid if $L_?$ is in the intersection of at least $n-d+1$ of the Cand_{L_i}, i.e., if

$$\exists I \subseteq \{1, \ldots, n\} : |I| \geq n-d+1 \quad \text{and} \quad L_? \in \cap_{i \in I} \mathsf{Cand}_{L_i} \ . \tag{8.9}$$

Example 8.2.10 From the set of extended clauses

$$\{X_1 + X_2 + X_3 \geq 2, X_1 + X_2 + X_4 \geq 2, X_1 + X_2 + X_5 \geq 2, X_1 + X_3 + X_4 \geq 2\} \tag{8.10}$$

we can derive the simplifying extended clause $X_1 + X_2 + X_3 + X_4 \geq 3$. While trying to eliminate $X_1 + X_2 + X_3 \geq 2$, we obtain the following Cand_{L_i}.

$$\mathsf{Cand}_{X_1} = \emptyset \qquad \mathsf{Cand}_{X_2} = \{X_4\} \qquad \mathsf{Cand}_{X_3} = \{X_4, X_5\} \tag{8.11}$$

Condition (8.8) does not apply but (8.9) does, because $n-d+1 = 2$ and X_4 is in the intersection of two of the Cand_{L_i}.

8.2.4 Temporary Diagonal Sums

We again consider Example 8.2.10, but we replace $X_1 + X_2 + X_4 \geq 2$ by $X_2 + X_4 \geq 1$. Therefore, Cand_{X_3} becomes $\{X_5\}$, and we can no longer derive the extended clause $X_1 + X_2 + X_3 + X_4 \geq 3$. Obviously, it would be possible if S contains $C := X_1 + X_2 + X_4 \geq 2$ or if $S \models C$. One approach to approximate $\models$ is to check whether C is a diagonal sum. By applying Definition 8.2.1 we see that C is a diagonal sum (and therefore valid) if $X_1 + X_2 \geq 1$, $X_1 + X_4 \geq 1$ and $X_2 + X_4 \geq 1$ are dominated by extended clauses in S. Since this is the case, we derive C and therefore $X_1 + X_2 + X_3 + X_4 \geq 3$. In other words, we enlarge the set Cand_{X_3} with the literal X_4 by recursively searching for a diagonal sum $L_? + X_2 + X_4 \geq 2$. In some sense we *temporarily* add the father $X_2 + X_4 \geq 1$ to the set of extended clauses, eliminate it with a simplifying diagonal sum, and use the computed candidates to construct the originally sought simplifying diagonal sum. In the following we generalize the idea of recursively searching for simplifying diagonal sums for the fathers of a simplifying diagonal sum.

Throughout this section let S be a set of extended clauses and $L \geq d \in S$ an extended clause for which we search a simplifying diagonal sum $L_? + L \geq d+1$. Let M always denote a subset of L. We call an extended clause of the form $L_? + L \setminus M \geq d + 1 - |M|$ also a *goal*. We denote by $\theta(M) := \{M \cup \{L_i\} \mid L_i \in L \setminus M\}$ the set of all sets of literals containing M and one more literal of L. We denote by

$$\mathsf{Cand}_M := \{L_? \mid C \models L_? + L \setminus M \geq d + 1 - |M| \text{ for a } C \in S\} \qquad (8.12)$$

the set of literals $L_?$ for which $L_? + L \setminus M \geq d + 1 - |M|$ is dominated by an extended clause in S. We reformulate Proposition 8.2.9 to be suitable for arbitrary goals.

Lemma 8.2.11 *The extended clause $L_? + L \setminus M \geq d + 1 - |M|$ is valid if*

$$\exists \Theta \subseteq \theta(M) : |\Theta| \geq n - d + 1 \qquad \text{and} \qquad L_? \in \cap_{M' \in \Theta} \mathsf{Cand}_{M'} \ . \qquad (8.13)$$

Proof. Note that $|L \setminus M| - (d - |M|) + 1 = |L| - d + 1 = n - d + 1$, because $M \subset L$. But then Proposition 8.2.9 applies, since we have (8.13). □

We say that we *expand* a goal $L_? + L \setminus M \geq d + 1 - |M|$ if we calculate all $L_?$ satisfying (8.13) and add them to Cand_M. In the example we obtained the new candidate X_4 for $L_?$ in $\mathsf{Cand}_{\{X_3\}}$ by expanding $L_? + X_1 + X_2 \geq 2$. Next, we define the set Cand_M^k of literals $L_?$ obtained by expanding k times $L_? + L \setminus M \geq d + 1 - |M|$. Cand_M^k is the set of literals such that $L_? + L \setminus M \geq d + 1 - |M|$ is valid w.r.t. S in the kth expansion. For $k = 0$ we define $\mathsf{Cand}_M^0 := \mathsf{Cand}_M$. For $k \geq 1$ we recursively define

$$\mathsf{Cand}_M^k := \{L_? \mid \exists \Theta \subseteq \theta(M) : |\Theta| \geq n - d + 1 \text{ and } L_? \in \cap_{M' \in \Theta} \mathsf{Cand}_{M'}^{k-1}\} \ . \tag{8.14}$$

By construction we see that $L_? + L \geq d+1$ is valid if there exists a k such that $L_? \in \mathsf{Cand}_\emptyset^k$. Note that (8.9) is equivalent to $L_? \in \mathsf{Cand}_\emptyset^1$. Obviously, $k \leq d$ because the degree must be greater than 1.

Lemma 8.2.12 *The number of goals in the k-th expansion of $L_? + L \geq d + 1$ is $\sum_{i=0}^{k} \binom{n}{i}$.*

Proof. While expanding a goal $L_? + L \geq d + 1$ we create all new goals where exactly 1 literal is missing, which are $\binom{n}{1} = n$ goals. Recursively expanding each of the new goals gives all goals, where exactly 2 literals are missing, which are $\binom{n}{2}$ goals. At level k we have $\binom{n}{k}$ new goals. The number of all goals is therefore $\sum_{i=0}^{k} \binom{n}{i}$. □

Next, we show that if we recursively search for candidates of $L_?$, then we only enlarge the set of candidates. For this purpose we first relate the expansion level and the number of literals that are missing in a goal.

Lemma 8.2.13 *If $M \subseteq M'$, then $\mathsf{Cand}_M^k \subseteq \mathsf{Cand}_{M'}^k$.*

Proof. Let $M \subseteq M'$ and suppose that $M' \in \theta(M)$. Let L_j be in Cand_M^k. We show that $L_j \in \mathsf{Cand}_{M'}^k$ by induction over k.

$k = 0$: $L_j + L \setminus M \geq d + 1 - |M|$ dominates $L_j + L \setminus M' \geq d + 1 - |M'|$.

$k - 1 \to k$: Because $L_j \in \mathsf{Cand}_M^k$, we have $\exists \Theta' \subseteq \theta(M) : |\Theta'| \geq n - d + 1$ and $L_? \in \bigcap_{N \in \Theta'} \mathsf{Cand}_N^{k-1}$. By induction hypothesis, we know that $\exists \Theta' \subseteq \theta(M') : |\Theta'| \geq n - d + 1$ and $L_? \in \bigcap_{N \in \Theta'} \mathsf{Cand}_N^{k-1}$ Therefore, $L_j \in \mathsf{Cand}_{M'}^k$. □

By induction $\mathsf{Cand}_M^k \subseteq \mathsf{Cand}_{M'}^k$ follows for any $M \subseteq M'$.

Lemma 8.2.14 *For all $l \geq k$, $\mathsf{Cand}_M^k \subseteq \mathsf{Cand}_M^l$ holds.*

Proof. It is sufficient to proof the lemma for $l = k+1$. Let $L_j \in \mathsf{Cand}_M^k$, then we know by Lemma 8.2.13 that for all M' in $\theta(M)$ we have $L_j \in \mathsf{Cand}_{M'}^k$. By definition (8.14) we have $L_j \in \mathsf{Cand}_M^{k+1}$. □

We show now that the candidates obtained in all goals of a level are necessarily candidates of the original goal.

Lemma 8.2.15 $\bigcap_{M:|M|=l} \mathsf{Cand}_M \subseteq \mathsf{Cand}_\emptyset^l$.

Proof. We know that if $L_j \in \mathsf{Cand}_M^k$ for all M with $|M| = l$, then $L_j \in \mathsf{Cand}_M^{k+1}$ for all M with $|M| = l-1$ by repeated application of definition (8.14). So if $L_j \in \bigcap_{M:|M|=l} \mathsf{Cand}_M^0$, then we have $L_j \in \bigcap_{M:|M|=l-1} \mathsf{Cand}_M^1$ and by induction $L_j \in \bigcap_{M:|M|=0} \mathsf{Cand}_M^l = \mathsf{Cand}_\emptyset^l$. □

Note that the converse does not necessarily hold. It is only true if we consider the case $|M| = d$. The idea is to restrict the number of levels that need to be expanded. For this purpose we need a technical lemma.

Lemma 8.2.16 *Let N be $\{1,\ldots,n\}$, $n \geq 2$ and let M be a subset of N. Let P be a property of subsets of N such that if $P(M)$ holds, then $P(M')$ holds for at least $t+1$ subsets M' of M with $|M'| = |M| - 1$ and $1 \leq t \leq n-1$. Now if $P(N)$ holds, then $P(M)$ holds for all subsets M of N with $|M| = t$.*

Proof. We show the lemma by induction on n.

$n = 2$: obvious.

$n-1 \to n$: If $t = n-1$ the lemma trivially holds, so assume $t < n-1$. Suppose that $P(N)$ holds. Then by definition we have $t+1$ different subsets N' of N such that $|N'| = n-1$ and $P(N')$ holds. Let N_i be $N \setminus \{i\}$ and w.l.o.g. suppose that $P(N_i)$ for all $1 \leq i \leq t+1$. By induction hypothesis we have for all $1 \leq i \leq t+1$ that $P(M_i)$ holds for all subsets M_i of N_i with $|M_i| = t$. It remains to show that every subset M of N such that $|M| = t$ is a subset of one of the N_i. Suppose otherwise, then there is $M \subset N$ that is not a subset of any of the N_i. But then for all N_i we derive that M must contain i and therefore $\{1,\ldots,t+1\} \subseteq M$, which contradicts $|M| = t$. □

We use Lemma 8.2.16 for showing that there is a literal for which we have a simplifying diagonal sum, obtained from any expansion level, if and only if this literal is in all the goals of the maximal expansion level.

Proposition 8.2.17 *There exists an expansion level k such that $\mathsf{Cand}_\emptyset^k$ is not empty if and only if $\bigcap_{M:|M|=d} \mathsf{Cand}_M$ is not empty.*

Proof. If $\bigcap_{M:|M|=d} \mathsf{Cand}_M$ is not empty, then by Lemma 8.2.15 we know that $\mathsf{Cand}_\emptyset^d$ is not empty. If $\mathsf{Cand}_\emptyset^k$ is not empty, then by Lemma 8.2.14 we know that $\mathsf{Cand}_\emptyset^d$ is not empty, since $k \leq d$. Let L_j be in $\mathsf{Cand}_\emptyset^d$. It remains to show that L_j in Cand_M for all M with $|M| = d$. For this we use Lemma 8.2.16. We know by Definition (8.14) that if $L_j \in \mathsf{Cand}_M$, then $L_j \in Cand_{M'}$ for $n-d+1$ sets $M' \in \Theta(M)$. We associate with Cand_M^k the set of indices of the literals in the corresponding goal. We associate with $Cand_\emptyset^d$ the set $\{1,\ldots,n\}$ and with the Cand_M the subsets of $\{1,\ldots,n\}$ with cardinality $n-d$. We associate with $n-d$ the value t and with the property $P(M)$ the property $L_i \in \mathsf{Cand}_M^{d-|M|}$, then the proposition follows. □

With Proposition 8.2.17 we can restrict the number of possible goals to be at most

$$\sum_{i=0}^{\min(n-d,d)} \binom{n}{i} .$$

Note that $\binom{n}{d} = \binom{n}{n-d}$ and each level k of a valid k-expansion for $k \geq n-d$ has more goals than the level d of the d-expansion. Because of Proposition 8.2.17, we can replace each k-expansion by level d of the d-expansion and lose no possible candidates. In particular, this is useful for extended clauses with weakness 1, so-called *cliques*[2] (Definition 7.7.5).

Corollary 8.2.18 *Let $L_1 + \cdots + L_n \geq n-1$ be a clique. Then there exists a simplifying diagonal sum $L_? + L_1 + \cdots + L_n \geq n$ if and only if one of the following equivalent conditions holds.*

- *For each $M \subset \{1,\ldots,n\}$ and $|M| = n-1$ we have $L_? \in \mathsf{Cand}_M$.*
- *For all $1 \leq i \leq n$ there is a $L_?$ such that the extended clause $L_? + L_i \geq 1$ is dominated by an extended clause in S.*

[2] In 0-1 integer programming *cliques* are usually written as $\leq$-inequalities and, then, are of the form $L_1 + \cdots + L_n \leq 1$.

Note that a simplifying diagonal sum obtained by a k-expansion is not necessarily a rank 1 cut, but may be a rank k cut.

8.2.5 Simplifying Clausal Satisfiability Problems

We apply the presented simplification algorithm on some combinatorial satisfiability problems and solve the 0-1 integer formulation of the simplified and the original problem with a linear programming based branch-and-bound algorithm. We use the commercial, state-of-the-art, mixed integer solver CPLEX.

We present some details of the implemented prototype. Given a set of extended clauses S, we try to find a simplifying diagonal sum and then simplify the problem. This is continued until no simplifying diagonal sum can be found. Searching for a simplifying diagonal sum is done incrementally, that is computed values for the different Cand_M in the different goals of the different extended clauses in S are not recomputed, but stored and reused after a successful search. As strategy, extended clauses having smaller weakness are preferred, because

- their simplifying diagonal sums are likely to dominate more other extended clauses.
- their maximal number of goals is smaller in the average case.
- the set of extended clauses that can contribute to the diagonal sum is smaller.

Note that the weakness of an extended clause is related to the length of classical clauses. If there are several extended clauses with the same weakness, then extended clauses with greater degree are preferred. We try to find a diagonal sum by $(k+1)$-expansion only if no diagonal sum is found for all extended clauses by k-expansion. We use Proposition 8.2.17 to restrict the maximal number of goals. If no further diagonal sum is found the simplified problem is given to CPLEX.

The examples, taken from [MR91], are combinatorial problems formulated as satisfiability problems. The "ulm..."-problems include n-queens problems, the "hole..."-problems are pigeon hole problems and the "nod..."-problems are

Table 8.1 **FAWULM (combinatorial problems (dataset 2/7))**

Name	Vars	Clauses	SDS	SClauses	CPLEX	Simp	SCPLEX
ulmbc024	24	164	60	38	0.15	0.52	0.02
ulmbc040	40	340	126	69	0.32	1.98	0.03
ulmbc060	60	610	218	108	3.22	5.40	0.35
ulmbc084	84	994	340	158	1.62	12.90	0.40
ulmbp048*	48	99	20	29	6.87	0.28	0.02
ulmbp070*	70	161	30	41	49.15	0.54	0.03
ulmbp096*	96	244	42	55	364.23	1.28	0.05
ulmbp126*	126	351	56	71	>1000	2.30	0.07
ulmbp160*	160	485	72	89	>1000	4.18	0.10
ulmbp198*	198	649	90	109	>1000	7.21	0.12
ulmbs040	40	70	15	25	0.07	0.14	0.02
ulmbs060	60	120	24	36	0.95	0.36	0.03
ulmbs084	84	189	35	49	6.70	0.78	0.03
ulmbs112	112	280	48	64	89.70	1.50	0.07
ulmbs144	144	396	63	81	0.45	2.94	0.08
ulmbs180	180	540	80	100	0.65	6.30	0.10
hole6*	42	133	30	13	25.02	0.38	0.02
hole7*	56	204	42	15	306.90	0.87	0.03
hole8*	72	297	56	17	>1000	1.71	0.02
hole9*	90	415	72	19	>1000	3.44	0.07
hole10*	110	561	90	21	>1000	5.86	0.08
hole20*	420	4221	380	41	>1000	291.30	0.10
nod5col4*	40	130	55	45	2.20	0.70	0.22
nod6col4*	60	255	97	64	12.68	2.21	0.05
nod6col5	75	315	130	87	0.27	3.40	0.03
nod7col4*	84	441	154	91	86.98	6.25	0.12
nod7col6*	126	651	225	120	>1000	11.91	797.42

graph-coloring problems. Table 8.1 is read as follows. In the first column the name of the problem is given, followed by $*$ if the problem is unsatisfiable. The second column is the number of boolean variables in the original problem followed by the number of classical clauses. In column "SDS" we count the number of simplification steps and in column "SClauses" we count the number of extended clauses in the simplified problem. In column "Simp" we find the time needed by the simplification algorithm prototyped in PROLOG. In column "CPLEX" we find the time CPLEX needs to solve the original problem. In the last column "SCPLEX" we find the time CPLEX needs to solve the simplified problem. All examples are run on a SPARC-10/31, all times are user-CPU-seconds. The presented preprocessing method behaves well on combinatorial satisfiability problems.

Example 8.2.19 The presented algorithm solves hard instances of the pigeon hole problems [Hak85] in sub-exponential time, see Table 8.1 (hole...). The problem is to fit m pigeons into n pigeonholes. Each pigeonhole may hold at most one pigeon, each pigeon must be placed into some hole. Let P_{ij} denote whether to place pigeon i in hole j. The problem is described by the following extended clauses. Each pigeon must be placed in one hole:

$$\sum_{j=1}^{n} P_{ij} \geq 1 \qquad \text{for all } 1 \leq i \leq m \tag{8.15}$$

Each hole can hold at most one pigeon:

$$\sum_{i=1}^{m} P_{ij} \leq 1 \quad (\equiv \sum_{i=1}^{m} \overline{P_{ij}} \geq m-1) \qquad \text{for all } 1 \leq j \leq n \tag{8.16}$$

Hard instances are obtained for $m = n + 1$. Using classical clauses the latter constraints are

$$\overline{P_{i_1 j}} + \overline{P_{i_2 j}} \geq 1 \qquad \text{for all } 1 \leq j \leq n, 1 \leq i_1 < i_2 \leq m\ . \tag{8.17}$$

Our simplification method deletes all classical clauses of (8.17) and replaces them with (8.16) by just generating simplifying clique inequalities. No more simplifications apply. For $m = n + 1$ the sum of all extended clauses in (8.15) and (8.16) is $0 \geq 1$. Therefore, the linear programming relaxation is unsatisfiable. The number of simplification steps is bounded by $O(n^3)$ like in [CCT87]. For each clique simplification we have at most n goals. Therefore, our simplification algorithm in combination with a linear programming solver solves the pigeon hole problems in polynomial time.

The main point here is that we have found a short proof of the pigeon hole formulae *automatically* with our simplification algorithm. So we exploit the theoretical power of cutting planes [CCT87] in a practical framework.

8.3 SIMPLIFYING RESOLVENTS

The resolution rule of *Generalized Resolution* [Hoo92a] is just classical resolution. We have presented in Section 7.8 a stronger inference rule that replaces resolution and allows us to generate immediately the strongest extended clauses derivable from two given extended clauses.

Restricting resolution to be simplifying turns out to be easy for classical resolution, as presented in Section 8.3.1. Since classical resolution only generates extended clauses with right-hand side 1, no extended clauses with right-hand side greater than 1 can be eliminated. In Section 8.3.2 we restrict the enhanced resolution rule of Section 7.8 to be simplifying. Unfortunately, searching simplifying resolvents with the enhanced resolution rule shows to be computationally expensive, since there is no way to restrict significantly the search for fathers of the simplifying resolvent. We overcome the problem by filtering out cases that are not promising.

8.3.1 Simplifying Resolvents for Classical Clauses

We say that a set of classical clauses S has a *simplifying resolvent* if there exists a classical clause $L \geq 1 \in S$ and a resolvent $L' \geq 1$ of two classical clauses in S such that $L' \subset L$. Note that $\mathsf{Ext}(S) = \mathsf{Ext}((S \setminus \{L \geq 1\}) \cup \{L' \geq 1\})$ and, since $L' \geq 1 \prec L \geq 1$, we say we have simplified S when we replace S by $(S \setminus \{L \geq 1\}) \cup \{L' \geq 1\}$.

For determining a simplifying resolvent we could compute all possible resolvents and then check whether there is one strictly dominating a classical clause in S, which is too inefficient. As done for the diagonal sum rule, we try to construct for each classical clause in S a resolvent of two classical clauses in S strictly dominating it.

Let $L \geq 1 \in S$ be a classical clause for which we search a simplifying resolvent $L' \geq 1$ with $L' \subset L$. First, we restate the resolution rule. Two classical clauses have a resolvent if and only if there is exactly one Boolean variable X_i that occurs positively in one classical clause and negatively in the other. Therefore, two classical clauses admitting a resolvent are of the form

$$M \cup \{X_i\} \geq 1 \tag{8.18}$$

$$N \cup \{\overline{X_i}\} \geq 1 \tag{8.19}$$

with $M \cap \overline{N} = \emptyset$. The resolvent of (8.18) and (8.19) is then the classical clause $M \cup N \geq 1$. Since we search for a resolvent strictly dominating $L \geq 1$, we know that $M \cup N \subset L$ needs to hold.

First, suppose that either (8.18) or (8.19) is $L \geq 1$. We then have a simplifying classical resolvent strictly dominating $L \geq 1$ if and only if there exists a classical clause $M \cup \{\overline{L_i}\} \geq 1$ in S with $L_i \in L$ and $M \subset L$.

Next, suppose that neither (8.18) nor (8.19) is $L \geq 1$. Then there must be two classical clauses (8.18) and (8.19) in S such that neither X_i nor $\overline{X_i}$ occurs in L, since S is in normal form (Assumption 8.0.1), and $M \cup N \subset L$. Thus, it is sufficient to consider only classical clauses in S that contain exactly one literal that does not occur in L. Note that each classical clause in S that is different from $L \geq 1$ contains at least one literal that does not occur in L, since S is in normal form (Assumption 8.0.1). In the worst case searching a simplifying resolvent for a specific clause $L \geq 1$ is not better than just generating all resolvents and checking domination afterwards. But, in practice, the restrition that each clause that need to be considered must have exactly one literal that does not occur in L means that we can easily, i.e., in linear time w.r.t. the number of clauses, eliminate a large fraction of the candidates. Only for the remaining candidates we have to do a pairwise, i.e., quadratic, check.

We obtain the following algorithm that returns a simplifying resolvent for a classical clause $L \geq 1 \in S$ if one exists, and returns $\perp$ otherwise.

Algorithm 8.3.1 *Simplifying Resolvent for Classical Clauses*

```
class_simpres(L ≥ 1, S)
    S' := {M ∪ {L_i} ≥ 1 | M ∪ {L_i} ≥ 1 ∈ S, M ⊂ L and L_i ∉ L}
    if ∃M ∪ {L_i} ≥ 1 ∈ S' : ‾L_i ∈ L then
        return L \ {‾L_i} ≥ 1
    endif
    forall M ∪ {X_j} ≥ 1 ∈ S' : X_j ∉ L do
        if ∃N ∪ {‾X_j} ≥ 1 ∈ S' : M ∪ N ⊂ L then
            return M ∪ N ≥ 1
        endif
    endforall
    return ⊥
end class_simpres
```

We illustrate the simplification of a set of classical clauses with help of the resolution rule by an example.

Example 8.3.2 Let S be

$$\{X_1+X_2+X_3 \geq 1, \overline{X_3}+X_4 \geq 1, X_1+X_2+X_4+X_5 \geq 1, \overline{X_1}+X_2+X_4+X_6 \geq 1\} \tag{8.20}$$

and let us search a simplifying classical resolvent strictly dominating $X_1+X_2+X_4+X_5 \geq 1 \in S$ following Algorithm 8.3.1. We start calculating S' and obtain

$$\{X_1 + X_2 + X_3 \geq 1, \overline{X_3} + X_4 \geq 1\} .$$

Since there is no simplifying resolvent where $X_1 + X_2 + X_4 + X_5 \geq 1$ is one of the fathers, we search for two classical clauses in S' whose resolvent strictly dominates $X_1+X_2+X_4+X_5 \geq 1$. We select $X_1+X_2+X_3 \geq 1$ containing the positive literal X_3 that does not occur in $\{X_1, X_2, X_4, X_5\}$ and resolve with a classical clause in S' containing $\overline{X_3}$, i.e., with $\overline{X_3} + X_4 \geq 1$. We obtain the resolvent

$$X_1 + X_2 + X_4 \geq 1 ,$$

which indeed strictly dominates $X_1+X_2+X_4+X_5 \geq 1$, and thus we can replace the weaker classical clause $X_1+X_2+X_4+X_5 \geq 1$ by $X_1+X_2+X_4 \geq 1$. Next, we try to find a classical clause strictly dominating $\overline{X_1} + X_2 + X_4 + X_6 \geq 1$. The only classical clause in the newly calculated S' is the previously generated $X_1+X_2+X_4 \geq 1$, which contains a negated literal of $\{\overline{X_1}, X_2, X_4, X_6\}$, namely X_1. We obtain the simplifying resolvent $X_2+X_4+X_6 \geq 1$, strictly dominating $\overline{X_1}+X_2+X_4+X_6 \geq 1$ by resolving $X_1+X_2+X_4 \geq 1$ and $\overline{X_1}+X_2+X_4+X_6 \geq 1$. No further simplifying resolvents can be found. The simplified set of classical clauses equivalent to (8.20) is

$$\{X_1+X_2+X_3 \geq 1, \overline{X_3}+X_4 \geq 1, X_1+X_2+X_4 \geq 1, X_2+X_4+X_6 \geq 1\} . \tag{8.21}$$

8.3.2 Simplifying Resolvents for Extended Clauses

A set of extended clauses S has a *simplifying resolvent* if there is an extended clause $L \geq d \in S$ such that there is an extended clause in $\mathsf{SCP}(\mathsf{SRed}(M + N \geq d_1 + d_2))$, strictly dominating $L \geq d$, where $M \geq d_1$ and $N \geq d_2$ are two extended clauses in S (see Proposition 7.8.2). Checking whether a set of extended clauses has a simplifying resolvent is unfortunately a computationally expensive task. For a complete check we have to build all possible sums of all possible pairs of two extended clauses in S (which is of $O(|S|^2)$) and transform them into a set of equivalent extended clauses with the method of Chapter 7.

Note that the transformation itself may produce several extended clauses per pair. For each extended clause in S and each generated extended clause, we have to check domination, which is at least of $O(|S|^3)$. This approach is far too expensive.

As for classical clauses the idea is to select a specific extended clause $L \geq d \in S$ and to try to construct a simplifying extended clause strictly dominating $L \geq d$. Unlike for classical clauses we have found no way to restrict significantly the set of possible candidates in advance.

Example 8.3.3 [a] Let $X_1+X_2+X_3+X_4 \geq 2 \in S$ be the extended clause for which we search a simplifying resolvent. Suppose that the extended clauses $X_5+X_6+X_7 \geq 2$ and $X_1+X_2+X_3+\overline{X_5}+\overline{X_6}+\overline{X_7} \geq 3$ are in S. Their sum is the simplifying resolvent $X_1 + X_2 + X_3 \geq 2$, although the extended clause $X_5+X_6+X_7 \geq 2$ contains no literal also occurring in the extended clause to be dominated. Even worse, if S contains the extended clauses $X_5+X_6+X_7 \geq 2$ and $\overline{X_5}+\overline{X_6}+\overline{X_7} \geq 2$ whose sum is $\perp$. Then all extended clauses are dominated, although in both extended clauses none of the literals in the extended clause to be dominated needs to occur.

[b] Let $X_1 + X_2 + X_3 + X_4 + X_5 + X_6 + X_7 + X_8 + X_9 \geq 5 \in S$ be the extended clause for which we search a simplifying resolvent. Suppose that the two extended clauses $X_1+X_2+X_3+X_4 \geq 3$ and $X_5+X_6+X_7+X_8 \geq 3$ are in S. Their sum is the simplifying resolvent $X_1 + X_2 + X_3 + X_4 + X_5 + X_6 + X_7 + X_8 \geq 6$, although both extended clauses have a smaller right-hand side than the one to be dominated.

[c] Let $X_1 + X_2 + X_3 + X_4 \geq 1 \in S$ be the extended clause for which we search a simplifying resolvent. Suppose that the two extended clauses $X_1+X_2+X_3+\overline{X_4}+\overline{X_5}+\overline{X_6} \geq 2$ and $X_1+X_2+X_3+X_4+X_5+X_6 \geq 2$ are in S. Their sum is $2 \cdot X_1 + 2 \cdot X_2 + 2 \cdot X_3 \geq 1$, which dominates the simplifying resolvent $X_1 + X_2 + X_3 \geq 1$, although both extended clauses have a greater weakness than the one to be dominated.

Example 8.3.3 indicates that there is no easy check of whether or not an extended clause summed up with another extended clause strictly dominates a given extended clause. Hence, we cannot filter out a large amount of extended clauses in advance in order to reduce the computational effort. Therefore, we restrict ourselves to special cases, where reducing the number of candidates is possible. More precisely, we do not search for simplifying resolvents $L' \geq d'$ strictly dominating $L \geq d$ with $L' \supseteq L$ and $d' > d$. Note that some of these simplifying extended clauses are already generated by the diagonal sum rule as in Section 8.2. We search for simplifying resolvents $L' \geq d'$ with $L' \subset L$ and $d' \geq d$.

Input Refutation

Given an extended clause $L \geq d$ in a set of extended clauses S, the goal is to find an extended clause

$$L \setminus \{L_i\} \geq d \ , \tag{8.22}$$

dominated by S with $L_i \in L$. We know that (8.22) is dominated by S if and only if the negation of (8.22),

$$\overline{L} \setminus \{\overline{L_i}\} \geq |L| - d \ , \tag{8.23}$$

together with S is unsatisfiable. In the following we present a method that will detect unsatisfiability of $S \cup \{\overline{L} \setminus \{\overline{L_i}\} \geq |L| - d\}$ by an input resolution like method, where we allow at most two steps in order to detect unsatisfiability. One step in the input resolution method means adding one extended clause starting with (8.23), plus some additional checks whether $\perp$ is derivable. First, we allow only one step and show that we get an input refutation if $L \geq d$ together with one extended clause dominates (8.22). The method is then expanded to two step input resolution, and we show that there is an input refutation if there are two extended clauses in S whose sum dominates (8.22). Furthermore, we present a sequential algorithm for two step input resolution that allows one to filter out extended clauses that are unlike to lead to $\perp$. We reduce the computational effort with most simplifying resolvents still being found.

One Step Input Refutation

The goal is to obtain $\perp$ starting with (8.23) by considering one extended clause in S. Note that the resolvent of $L \geq d$ and (8.23) is $L_i \geq 1$. Therefore, we already have shown unsatisfiability of $S \cup \{\overline{L} \setminus \{\overline{L_i}\} \geq |L| - d\}$ if we can derive $\overline{L_i} \geq 1$. We try to obtain $\overline{L_i} \geq 1$ by input resolution starting with the extended clause (8.23).

Example 8.3.4 Let $X_1 + X_2 + X_3 + X_4 + X_5 + X_6 + X_7 \geq 3$ be the extended clause in S for which we search a simplifying resolvent and suppose that S contains the extended clause $X_1 + X_2 + X_3 + \overline{X_5} + \overline{X_6} \geq 4$. The sum of both extended clauses is $2 \cdot X_1 + 2 \cdot X_2 + 2 \cdot X_3 + X_4 + X_7 \geq 5$, which dominates $X_1 + X_2 + X_3 + X_4 + X_7 \geq 3$, and thus, for example, $X_1 + X_2 + X_3 + X_4 + X_6 + X_7 \geq 3$, a simplifying extended clause which strictly dominates $X_1 + X_2 +$

$X_3 + X_4 + X_5 + X_6 + X_7 \geq 3$. Next, we show with help of an input refutation that $X_1 + X_2 + X_3 + X_4 + X_6 + X_7 \geq 3$ is dominated by S. The literal that is missing from $\{X_1, X_2, X_3, X_4, X_5, X_6, X_7\}$ is X_5. We start resolving on the negation of $X_1 + X_2 + X_3 + X_4 + X_6 + X_7 \geq 3$, i.e.,

$$\overline{X_1} + \overline{X_2} + \overline{X_3} + \overline{X_4} + \overline{X_6} + \overline{X_7} \geq 4 \ , \tag{8.24}$$

and know that unsatisfiability is detected if $\overline{X_5} \geq 1$ follows. We sum up (8.24) and $X_1 + X_2 + X_3 + \overline{X_5} + \overline{X_6} \geq 4$, and obtain the linear pseudo-Boolean inequality

$$\overline{X_4} + \overline{X_5} + 2 \cdot \overline{X_6} + \overline{X_7} \geq 5 \ , \tag{8.25}$$

which fixes all literals in it and thus dominates $\overline{X_5} \geq 1$. Together with $X_5 \geq 1$ we obtain $\perp$ and thus have derived unsatisfiability.

Typically, we do not know in advance which literal $L_i \in L$ can be eliminated. Therefore, the problem is whether one of the extended clauses represented by the compact extended clause set

$$L^{|L|-1} \geq d \ , \tag{8.26}$$

(see Definition 7.4.3) is valid w.r.t. S. This is the case if and only if one of the extended clauses of the compact extended clause set

$$\overline{L}^{|L|-1} \geq |L| - d \ , \tag{8.27}$$

i.e., the negation of (8.26), together with S is unsatisfiable. Resolving $L \geq d$ with (8.27) yields the compact extended clause set

$$L^1 \geq 1 \ ,$$

which is $L_i \geq 1$ if we have decided which literal L_i we eliminate, i.e., which of the extended clauses represented by (8.26) we select.

Let us add an extended clause containing a literal L_i to a compact extended clause set, where $\overline{L_i} \in S_j^{k_j}$ for some set of literals $S_j^{k_j}$ in the compact representation. We consider only expansions, where we select $\overline{L_i}$ from $S_j^{k_j}$. We obtain a valid compact extended clause if we replace $S_j^{k_j}$ by $\overline{L_i}+(S_j \setminus \{\overline{L_j}\})^{k_j-1}$. The sum of the extended clause and the valid compact extended clause set then contains the subterm $L_i + \overline{L_i} + (S_j \setminus \{\overline{L_j}\})^{k_j-1}$, which simplifies to $1 + (S_j \setminus \{\overline{L_j}\})^{k_j-1}$, where 1 can be brought to the right-hand side. Performing this simplification for all literals if possible yields the *resolvent* of a compact extended clause set and an extended clause.

We start the refutation sequence with (8.27). Note that we can delete ay any time some literal $\overline{L_i} \in \overline{L}$, and have shown unsatisfiability if $\overline{L_i} \geq 1$ is dominated.

Example 8.3.5 Let us again consider Example 8.3.4. We start resolving on the negation of $\{X_1, X_2, X_3, X_4, X_5, X_6, X_7\}^6 \geq 3$, namely

$$\{\overline{X_1} + \overline{X_2} + \overline{X_3} + \overline{X_4} + \overline{X_5} + \overline{X_6} + \overline{X_7}\}^6 \geq 4 \ , \tag{8.28}$$

and add the extended clause $X_1 + X_2 + X_3 + \overline{X_5} + \overline{X_6} \geq 4$ yielding

$$\{\overline{X_4} + \overline{X_5} + \overline{X_6} + \overline{X_7}\}^3 + \overline{X_5} + \overline{X_6} \geq 5 \ . \tag{8.29}$$

Since we have 5 literals in the left-hand side and the right-hand side is 5, we know that all literals need to be fixed. Because we have to derive $\overline{L_i} \geq 1$ for the literal that we delete, the literals in question are X_5 and X_6. We select X_5 and obtain $\overline{X_4} + \overline{X_7} + \overline{X_5} + 2 \cdot \overline{X_6} \geq 5$, which dominates $\overline{X_5} \geq 1$. Since $X_5 \geq 1$ is obtained by summing up $\overline{X_1} + \overline{X_2} + \overline{X_3} + \overline{X_4} + \overline{X_6} + \overline{X_7} \geq 4$ and $X_1 + X_2 + X_3 + X_4 + X_5 + X_6 + X_7 \geq 3$, we obtain $\perp$ and thus know that

$$X_1 + X_2 + X_3 + X_4 + X_6 + X_7 \geq 3$$

is dominated by S. Hence, we have a simplifying resolvent strictly dominating $X_1 + X_2 + X_3 + X_4 + X_5 + X_6 + X_7 \geq 3$.

Next, we give an Algorithm for detecting a simplifying resolvent of an extended clause $L \geq d \in S$ by an input refutation, where we resolve only with one additional extended clause in S.

Algorithm 8.3.6 *One Step Input Resolution*

```
onestep_simpres(L ≥ d, S)
    forall L' ≥ d' ∈ S do
        rhs := |L| − d + d' − |L ∩ L'|
        X := L̄ \ (L̄ ∩ L̄')
        Y := L' \ (L ∩ L')
        (* X^{|X|−1} + Y ≥ rhs *)
        (* L̄_i ≥ 1 is dominated (fixing) *)
        if |X ∩ Y| ≥ 1 ∧ |X| + |Y| − 1 ≤ rhs then
            L_i ∈ X̄ ∩ Ȳ
            return L \ {L_i} ≥ d
        endif
    endforall
    return ⊥
end onestep_simpres
```

The correctness of Algorithm 8.3.6 will be proved in Proposition 8.3.8. Algorithm 8.3.6 returns a simplifying resolvent for $L \geq d \in S$ if it finds one by resolving with one extended clause $L' \geq d' \in S$. We show that Algorithm 8.3.6 finds a simplifying resolvent for $L \geq d$ if and only if there is an extended clause $L' \geq d' \in S$ such that the linear pseudo-Boolean inequality $L + L' \geq d + d'$ dominates $L \setminus \{L_i\} \geq d$ for an $L_i \in L$. For this purpose we first investigate when a linear pseudo-Boolean inequality, obtained by summing up two extended clauses, dominates an extended clause.

The sum of two extended clauses is always of the form

$$2 \cdot L + L' \geq d \ , \tag{8.30}$$

where L is the set of literals occurring in both extended clauses, and L' is the set of literals whose variables occur once in both extended clauses. Literals that occur positively in one and negatively in the other extended clause are simplified to 1 and brought to the right-hand side. Next, we investigate conditions when (8.30) dominates the extended clause $L \cup L' \geq d'$.

Lemma 8.3.7 *The linear pseudo-Boolean inequality* $2 \cdot L + L' \geq d$ *dominates the extended clause* $L \cup L' \geq d'$ *if and only if*

$$\begin{array}{ll} d' \leq \lceil d/2 \rceil & \text{if } 2|L| \geq d \\ d' \leq d - |L| & \text{if } 2|L| < d \ . \end{array} \tag{8.31}$$

Proof. By Proposition 7.1.2 we know that $2 \cdot L + L' \geq d$ dominates the extended clause $L \cup L' \geq d'$ if and only if the right-hand side of the strongest extended clause $\mathsf{SCP}(2 \cdot L + L' \geq d)$ is greater than or equal to d'. It remains to show that

$$\beta := \begin{cases} \lceil d/2 \rceil & \text{if } 2|L| \geq d \\ d - |L| & \text{if } 2|L| < d \end{cases}$$

is the right-hand side of $\mathsf{SCP}(2 \cdot L + L' \geq d)$.

First, suppose that $2|L| \geq d$. Condition (7.4) for calculating β states that β is the smallest number of the largest coefficients such that their sum is greater or equal than d. Since the sum of the literals with coefficient 2 is already greater than d, we know that we have to sum up $\lceil d/2 \rceil$ times the coefficient 2.

Next, suppose that $2|L| < d$. Then we have to sum up all coefficients that are 2, i.e., $|L|$, plus as many times 1 as needed in order to reach d, i.e., $d - 2|L|$. Thus, we need to sum up at least $|L| + d - 2|L| = d - |L|$ coefficients. □

Note that a linear pseudo-Boolean inequality $2 \cdot L + 2 \cdot M + L' + M' \geq d$ dominates an extended clause $L + L' \geq d'$ if and only if its reduction $2 \cdot L + L' \geq d - 2|M| - |M'|$ dominates $L + L' \geq d'$.

Proposition 8.3.8 *If there exists $L \geq d \in S$ and $L' \geq d' \in S$ such that $L + L' \geq d + d'$ dominates a simplifying resolvent $L \setminus \{L_i\} \geq d$ with $L_i \in L$, then there is an input refutation, following Algorithm 8.3.6, for a simplifying resolvent of $L \geq d$.*

Proof. Let $L + L' \geq d + d'$ dominate a simplifying resolvent $L \setminus \{L_i\} \geq d$ with $L_i \in L$. Let $M := L \cap L'$, $N := L \cap \overline{L'}$, $Q := L \setminus (M \cup N)$, and $R := L' \setminus (M \cup \overline{N})$. Then $M \cup N \cup Q = L$, $M \cup \overline{N} \cup R = L'$, and $L + L' \geq d + d'$ is

$$2 \cdot M + Q + R \geq d + d' - |N| \ .$$

Since R and L are disjoint, we know that

$$2 \cdot M + Q \geq d + d' - |N| - |R| \tag{8.32}$$

dominates $L \setminus \{L_i\} \geq d$.

First, we show that $N \neq \emptyset$. Suppose otherwise, then $L = M \cup Q$, $L' = M \cup R$, and we know that (8.32), which is

$$2 \cdot M + Q \geq d + d' - |R| \ , \tag{8.33}$$

dominates $L \setminus \{L_i\} \geq d$. Since $L = M \cup Q$ we know that (8.33) dominates $L \geq d + 1$. From Lemma 8.3.7 we know that then

$$\begin{array}{ll} d + 1 \leq \lceil (d + d' - |R|)/2 \rceil & \text{if } 2|M| \geq d + d' - |R| \\ d + 1 \leq d + d' - |R| - |M| & \text{if } 2|M| < d + d' - |R| \ . \end{array}$$

Note that if $2|M| < d + d' - |R|$, then $d + 1 \leq d + d' - |R| - |M|$, which is equivalent to $|R| + |M| + 1 \leq d'$. But then $d' > |L'|$ and thus $L' \geq d'$ is $\perp$, which is not allowed. Therefore, we have $2|M| \geq d + d' - |R|$ and thus

$$d + 1 \leq \lceil (d + d' - |R|)/2 \rceil \ . \tag{8.34}$$

From (8.34) we have

$$\begin{array}{rcl} 2d + 2 & \leq & d + d' - |R| + 1 \\ d + 1 & \leq & d' - |R| \\ d + 1 + |R| & \leq & d' \ . \end{array}$$

Thus, $L' \geq d'$ dominates $M \cup R \geq d + 1 + |R|$, which in turn dominates its reduction $M \geq d + 1$. Therefore, $L' \geq d'$ dominates $L \geq d$, which contradicts the normal form condition. We conclude that $N \neq \emptyset$ and w.l.o.g. we have $L_i \in N$.

Therefore, $L \setminus \{L_i\} \geq d$ is dominated by (8.32) if and only if (8.32) dominates $M + Q \geq d$. From Lemma 8.3.7 we know that

$$\begin{array}{ll} d \leq \lceil (d + d' - |N| - |R|)/2 \rceil & \text{if } 2|M| \geq d + d' - |N| - |R| \\ d \leq d + d' - |N| - |R| - |M| & \text{if } 2|M| < d + d' - |N| - |R| \end{array}$$

holds. Note that if $2|M| < d + d' - |N| - |R|$, then $d \leq d + d' - |N| - |R| - |M|$, which is equivalent to $|N| + |R| + |M| \leq d'$. But then $d' \geq |L'|$, which is not allowed, since the weakness of $L' \geq d$ is less than or equal to 0. Thus, either unsatisfiability is detected or the literals in L' are fixed to 1. Therefore, we have $2|M| \geq d + d' - |N| - |R|$ and thus

$$d \leq \lceil (d + d' - |N| - |R|)/2 \rceil \ . \tag{8.35}$$

Next, we establish that $d = \lceil (d + d' - |N| - |R|)/2 \rceil$. Suppose otherwise, then we have

$$\begin{array}{rcl} d + 1 & \leq & \lceil (d + d' - |N| - |R|)/2 \rceil \\ 2d + 2 & \leq & d + d' - |N| - |R| + 1 \\ d + 1 & \leq & d' - |N| - |R| \\ d' & \geq & d + 1 + |N| + |R| \ . \end{array}$$

Therefore, $L' \geq d'$ dominates $M \cup \overline{N} \cup R \geq d+1+|N|+|R|$, since $L' = M \cup \overline{N} \cup R$, which dominates its reduction $M \geq d + 1$. But then $L' \geq d'$ dominates $L \geq d$, since $M \subseteq L$, which contradicts the normal form condition. We conclude that $d = \lceil (d + d' - |N| - |R|)/2 \rceil$.

Let us now investigate an input refutation starting with $\overline{M} \cup \overline{N}^{|N|-1} \cup \overline{Q} \geq |M| + |N| + |Q| - d$. We add $M \cup \overline{N} \cup R \geq d'$, which yields

$$\overline{N} + \overline{N}^{|N|-1} + \overline{Q} + R \geq |N| + |Q| - d + d' \ . \tag{8.36}$$

Since (8.36) needs to dominate $\overline{L_i} \geq 1$, where L_i is the literal that is missing from $\overline{N}$ in $\overline{N}^{|N|-1}$, it remains to show

$$2|N| - 1 + |Q| + |R| \leq |N| + |Q| - d + d' \ , \tag{8.37}$$

which simplifies to $|N|+|R|+d \leq d'+1$. We assume the converse,

$$|N|+|R|+d > d'+1 \ , \tag{8.38}$$

which will lead to a contradiction. From (8.38) we have $|N|+|R|+d \geq d'+2$. Since $d = \lceil (d+d'-|N|-|R|)/2 \rceil$, we derive

$$|N|+|R|+\lceil (d+d'-|N|-|R|)/2 \rceil \geq d'+2 \ ,$$

which implies

$$\begin{aligned} 2|N|+2|R|+d+d'-|N|-|R|+1 &\geq 2d'+4 \\ |N|+|R|+d &\geq d'+3 \\ d &\geq d'+3-|N|-|R| \ . \end{aligned}$$

Since $d \geq d'+3-|N|-|R|$, we know that $L \geq d$ dominates

$$M \cup N \cup R \geq d'+3-|N|-|R|$$

and thus dominates

$$M \geq d'+3 \ .$$

But then $L \geq d$ dominates $L' \geq d'$, which again violates the normal form condition.

Note that in Algorithm 8.3.6 X is $\overline{N} \cup \overline{Q}$, Y is $\overline{N} \cup R$, and rhs is $|N|+|Q|-d+d'$. Thus $|X|+|Y|-1 \leq rhs$ is equivalent to

$$2|N|+|R|+|Q|-1 \leq |N|+|Q|-d+d'$$

which is (8.37). Furthermore, $X \cap Y = \overline{N}$ and thus we also ensure that N is not empty. □

Therefore, Algorithm 8.3.6 is a complete method checking whether there is an extended clause $L' \geq d' \in S$, which dominates together with $L \geq d$ a simplifying resolvent $L \setminus \{L_i\} \geq d$. A more general method is to check whether there are two extended clauses in S whose sum dominates a simplifying resolvent for $L \geq d$. We present a similar input refutation method for that problem.

Two Step Input Refutation

Let $L \geq d \in S$ be an extended clause for which we want to check whether there are two extended clauses in S such that their sum dominates a simplifying resolvent $L \setminus \{L_i\} \geq d$. We present an input refutation method starting with $\overline{L}^{|L|-1} \geq |L|-d$, but allows two steps in order to derive unsatisfiability.

Example 8.3.9 Let the two extended clauses $X_1 + X_2 + X_5 + X_6 + X_7 \geq 2$ and $X_3 + X_4 + \overline{X_5} + \overline{X_6} + \overline{X_7} \geq 3$ be in S. Their sum $X_1 + X_2 + X_3 + X_4 \geq 2$ is a simplifying resolvent for $X_1 + X_2 + X_3 + X_4 + X_8 \geq 2 \in S$. We start an input refutation with $\{\overline{X_1}, \overline{X_2}, \overline{X_3}, \overline{X_4}, \overline{X_8}\}^4 \geq 3$, and add the two extended clauses $X_1 + X_2 + X_5 + X_6 + X_7 \geq 2$ and $X_3 + X_4 + \overline{X_5} + \overline{X_6} + \overline{X_7} \geq 3$ yielding $\{\overline{X_8}\}^0 \geq 1$, which is $\perp$.

We see that in a two step input refutation we may immediately obtain $\perp$. It is also possible to derive that some literals need to be fixed, and fixing these literals in $\overline{L}^{|L|-1} \geq |L| - d$ yields the desired $\perp$.

Example 8.3.10 Let $X_1 + X_2 + X_3 + X_4 + X_5 + X_8 + X_9 + X_{10} \geq 4$ and $X_1 + X_2 + X_3 + X_6 + X_7 + \overline{X_8} + \overline{X_9} + \overline{X_{10}} \geq 5$ be two extended clauses in S. Their sum $2 \cdot X_1 + 2 \cdot X_2 + 2 \cdot X_3 + X_4 + X_5 + X_6 + X_7 \geq 6$ dominates, for example, the extended clause $X_1 + X_2 + X_3 + X_4 + X_5 + X_6 \geq 3$, which is a simplifying resolvent for the extended clause $X_1 + X_2 + X_3 + X_4 + X_5 + X_6 + X_{11} \geq 3 \in S$. We start an input refutation with

$$\{\overline{X_1}, \overline{X_2}, \overline{X_3}, \overline{X_4}, \overline{X_5}, \overline{X_6}, \overline{X_{11}}\}^6 \geq 4 \ . \tag{8.39}$$

Adding both extended clauses yields $X_1 + X_2 + X_3 + X_7 + \{X_{11}\}^0 \geq 4$. Thus, the literals X_1, X_2, X_3 and X_7 need to be fixed to 1. The literal that needs to be eliminated is X_{11}. Fixing X_1, X_2 and X_3 to 1 in (8.39) yields $\{\overline{X_4}, \overline{X_5}, \overline{X_6}, \overline{X_{11}}\}^3 \geq 4$, which is $\perp$, and thus unsatisfiability is shown.

Unfortunately, these two cases are not sufficient. It is also possible that $\perp$ may only be obtained by adding again the initial compact extended clause set.

Example 8.3.11 Let $X_1 + X_2 + X_3 + X_4 + X_5 + X_6 + X_7 \geq 3$ and $X_1 + X_2 + X_3 + X_4 + X_5 + \overline{X_6} + X_8 \geq 3$ be two extended clauses in S. Their sum $2 \cdot X_1 + 2 \cdot X_2 + 2 \cdot X_3 + 2 \cdot X_4 + 2 \cdot X_5 + X_7 + X_8 \geq 5$ dominates the extended clause $X_1 + X_2 + X_3 + X_4 + X_5 + X_7 + X_8 \geq 3$, which is a simplifying resolvent for $X_1 + X_2 + X_3 + X_4 + X_5 + X_7 + X_8 + X_9 \geq 3 \in S$. We start an input refutation with

$$\{\overline{X_1}, \overline{X_2}, \overline{X_3}, \overline{X_4}, \overline{X_5}, \overline{X_7}, \overline{X_8}, \overline{X_9}\}^7 \geq 5 \ , \tag{8.40}$$

and add the two extended clauses yielding $X_1 + X_2 + X_3 + X_4 + X_5 + \{X_9\}^0 \geq 3$, which is neither $\perp$ nor fixes any literal. Note that the set of literals that occur in both extended clauses and in the extended clause to be dominated is relatively large, i.e., its cardinality is greater than or equal to the right-hand side. One way to eliminate these literals is to add again (8.40) yielding $\{\overline{X_7}, \overline{X_8}, \overline{X_9}\}^2 + \{X_9\}^0 \geq 3$, which is $\perp$, and thus we have shown unsatisfiability.

L	L_i	L_i	L_i	L_i	L_i	L_i	L_i	L_i	L_i				
L'	L_i	L_i	L_i	$\overline{L_i}$	$\overline{L_i}$	$\overline{L_i}$				L_i	L_i	L_i	
L''	L_i	$\overline{L_i}$		L_i	$\overline{L_i}$		L_i		$\overline{L_i}$	L_i	$\overline{L_i}$		L_i
Set	M	N	Q	P	$\overline{M'}, T$	$\overline{Q'}, T$	R	T	$\overline{R'}, T$	M'	N'	Q'	R'

Figure 8.1 Splitting literals

Based on these examples, we construct the following algorithm for detecting a simplifying resolvent of $L \geq d \in S$, where we resolve with two other extended clause in S. When we select $L' \geq d_1, L'' \geq d_2$, a pair of extended clauses in S, we build 9 disjoint subsets (M, M', N, N', Q, Q', R, R', T) of the literals occurring in the extended clauses $L \geq d$, $L' \geq d_1$ and $L'' \geq d_2$ according to Figure 8.1. The first three lines indicate the occurrence of a literal, resp. its negation, in the sets of literals L, L' and L''. The fourth line indicates in which set to put L_i. For example, the seventh column reads "If L_i occurs in L, $\overline{L_i}$ occurs in L', and neither L_i nor $\overline{L_i}$ occurs in L'', then put $\overline{L_i}$ in the set Q' and put L_i in the set T ". Figure 8.1 covers all possible cases exactly once. Note that $L = M \cup N \cup P \cup Q \cup R \cup T$, $L' = M \cup M' \cup N \cup \overline{P} \cup N' \cup Q \cup Q'$, and $L'' = M \cup M' \cup \overline{N} \cup P \cup \overline{N}' \cup R \cup R'$. This splitting enables us to express directly the combinations of the extended clauses and to reason about their properties like in Proposition 8.3.13.

Algorithm 8.3.12 *Two Step Input Resolution*

$\texttt{twostep\_simpres}(L \geq d, S)$

forall $\begin{array}{l} M + M' + N + \overline{P} + N' + Q + Q' \geq d_1, \\ M + M' + \overline{N} + P + \overline{N'} + R + R' \geq d_2 \end{array} \in S$ **do**

$\quad T := L \setminus (M \cup N \cup P \cup Q \cup R)$

$\quad$ **if** $N = P = T = \emptyset$ **return** $\perp$ **endif**

$\quad$ **if** $|M| + |N| + |P| \leq d_1 + d_2 - d - |N'| - 2|M'| - |Q'| - |R'| \quad \vee$

$\qquad (d < |M| + 1 \quad \wedge$

$\qquad |M| + |N| + |P| - 1 = d_1 + d_2 - d - |N'| - 2|M'| - |Q'| - |R') \quad \vee$

$\qquad (d \leq |M| - 1 \quad \wedge$

$\qquad 2d < d_1 + d_2 + |N| - |N'| - 2|M'| - |Q'| - |R'| + 2)$

$\quad$ **then**

$\qquad L_i \in N \cup P \cup T$

$\qquad$ **return** $L \setminus \{L_i\} \geq d$

$\quad$ **endif**

endforall
return $\bot$
end `twostep_simpres`

The first **if**-condition checks whether a simplifying resolvent is possible, i.e., whether a literal of L is missing in the sum of the two extended clauses. The next **if**-condition checks (a) whether the sum of the two extended clauses and $\overline{L}^{|L|-1} \geq |L| - d$ is $\bot$, (b) whether the sum of the two extended clauses and $\overline{L}^{|L|-1} \geq |L| - d$ forces fixing literals, and if yes whether fixing these literals in $\overline{L}^{|L|-1} \geq |L| - d$ yields $\bot$, and (c) whether $|M|$ is large enough such that the special case of Example 8.3.11 may occur. If yes, it is tested if the sum of the two extended clauses plus two times $\overline{L}^{|L|-1} \geq |L| - d$ yields $\bot$. It turns out that these three cases cover all possibilities.

Proposition 8.3.13 *If the sum $L' + L'' \geq d_1 + d_2$ of two extended clauses $L' \geq d_1$ and $L'' \geq d_2$ dominates an extended clause $L \setminus \{L_i\} \geq d$ with $L_i \in L$, but not $L \geq d + 1$, then there is an input refutation of $\overline{L}^{|L|-1} \geq |L| - d$ using these two extended clauses following Algorithm 8.3.12 and L_i is determined such that $L \setminus \{L_i\} \geq d$ is valid and strictly dominating $L \geq d$.*

Proof. The two extended clauses $L' \geq d_1$ and $L'' \geq d_2$ are of the form

$$M + M' + N + \overline{P} + N' + Q + Q' \geq d_1 \tag{8.41}$$
$$M + M' + \overline{N} + P + \overline{N'} + R + R' \geq d_2 \tag{8.42}$$

according to Figure 8.1. The sum of (8.41) and (8.42) is

$$2 \cdot M + 2 \cdot M' + Q + Q' + R + R' \geq d_1 + d_2 - |N| - |P| - |N'| \ . \tag{8.43}$$

Since (8.43) dominates $L \setminus \{L_i\} \geq d$, we know that the reduction

$$2 \cdot M + Q + R \geq d_1 + d_2 - |N| - |P| - |N'| - 2|M'| - |Q'| - |R'| \tag{8.44}$$

of (8.43) dominates $L \setminus \{L_i\} \geq d$. Note that $N \cup P \cup T \neq \emptyset$. Suppose otherwise, then $L = M \cup Q \cup R$ and thus we know that (8.44), which is

$$2 \cdot M + Q + R \geq d_1 + d_2 - |N'| - 2|M'| - |Q'| - |R'| \ , \tag{8.45}$$

dominates $M + Q + R \geq d + 1$, since it also dominates $L \setminus \{L_i\} \geq d$, which is $(M \cup Q \cup R) \setminus \{L_i\} \geq d$. But that case is excluded, thus, $N \cup P \cup T \neq \emptyset$. Therefore, we can safely suppose in the following that $L_i \in N \cup P \cup T$.

We start an input refutation with $\overline{L}^{|L|-1} \geq |L| - d$, which is

$$\overline{M} + \overline{Q} + \overline{R} + \{\overline{N} + \overline{P} + \overline{T}\}^{|N|+|P|+|T|-1} \geq |M| + |N| + |P| + |Q| + |R| + |T| - d \ . \tag{8.46}$$

We add the two extended clauses (8.41) and (8.42) to (8.46), and obtain

$$M + 2|M'| + Q' + R' + \{\overline{N} + \overline{P} + \overline{T}\}^{|N|+|P|+|T|-1} \geq d_1 + d_2 + |T| - d - |N'| \ , \tag{8.47}$$

which dominates its reduction

$$M + \{\overline{N} + \overline{P} + \overline{T}\}^{|N|+|P|+|T|-1} \geq d_1 + d_2 - d + |T| - |N'| - 2|M'| - |Q'| - |R'| \ . \tag{8.48}$$

We have shown unsatisfiability if

- inequality (8.48) is $\perp$, i.e., if

$$|M| + |N| + |P| \leq d_1 + d_2 - d - |N'| - 2|M'| - |Q'| - |R'| \ , \tag{8.49}$$

- or if the literals in $M + \{\overline{N} + \overline{P} + \overline{T}\}^{|N|+|P|+|T|-1}$ need to be fixed, and fixing these literals in (8.46) yields $\perp$, i.e., if

$$|M| + |N| + |P| - 1 = d_1 + d_2 - d - |N'| - 2|M'| - |Q'| - |R'| \tag{8.50}$$

  and

$$|\overline{Q}| + |\overline{R}| + |\overline{N}| + |\overline{P}| + |\overline{T}| - 1 < |M| + |N| + |P| + |Q| + |R| + |T| - d \ , \tag{8.51}$$

  where (8.51) simplifies to

$$d < |M| + 1 \ , \tag{8.52}$$

- or if $d \leq |M| - 1$ and the sum of $2 \cdot$ (8.46), (8.41) and (8.42) is $\perp$.

In the remaining part of the proof we investigate the sum of (8.41) and (8.42), which dominates $L \setminus \{L_i\} \geq d$. We show that always one of the previous three cases apply, which shows the proposition.

We know that (8.44) dominates $L \setminus \{L_i\} \geq d$ if and only if it dominates $M + Q + R \geq d$, since $L_i \in N \cup P \cup T$. From Lemma 8.3.7 we know that

$$\begin{aligned} d \leq\ & \lceil d_1 + d_2 - |N| - |P| - |N'| - 2|M'| - |Q'| - |R'|/2 \rceil \\ & \text{if } 2|M| \geq d_1 + d_2 - |N| - |P| - |N'| - 2|M'| - |Q'| - |R'| \\ d \leq\ & d_1 + d_2 - |N| - |P| - |N'| - 2|M'| - |Q'| - |R'| - |L| \\ & \text{if } 2|M| < d_1 + d_2 - |N| - |P| - |N'| - 2|M'| - |Q'| - |R'| \ . \end{aligned}$$

- First, suppose that $2|M| < d_1 + d_2 - |N| - |P| - |N'| - 2|M'| - |Q'| - |R'|$, and thus $d \leq d_1 + d_2 - |N| - |P| - |N'| - 2|M'| - |Q'| - |R'| - |L|$. Since $L = M \cup N \cup P \cup Q \cup R \cup T$, we have

$$d \leq d_1 + d_2 - 2|N| - 2|P| - |N'| - 2|M'| - |Q'| - |R'| - |M| - |Q| - |R| - |T| \; ,$$

and after a rearrangement we know that

$$\begin{array}{c} |M| + |N| + |P| + |T| \quad \leq \\ d_1 + d_2 - d - |N| - |P| - |N'| - 2|M'| - |Q'| - |R'| - |Q| - |R| \; . \end{array} \tag{8.53}$$

We add the valid inequality $-|T| \leq |N| + |P| + |Q| + |R|$ to (8.53) and obtain

$$|M| + |N| + |P| \leq d_1 + d_2 - d - |N'| - 2|M'| - |Q'| - |R'| \; ,$$

which is (8.49) and thus we have shown unsatisfiability.

- Next, suppose that $2|M| \geq d_1 + d_2 - |N| - |P| - |N'| - 2|M'| - |Q'| - |R'|$, thus $d \leq \lceil d_1 + d_2 - |N| - |P| - |N'| - 2|M'| - |Q'| - |R'|/2 \rceil$, and therefore

$$2 \cdot d \leq d_1 + d_2 - |N| - |P| - |N'| - 2|M'| - |Q'| - |R'| + 1 \; . \tag{8.54}$$

We assume that input refutation fails to exhibit unsatisfiability.
First, suppose that

$$|M| + |N| + |P| - 1 > d_1 + d_2 - d - |N'| - 2|M'| - |Q'| - |R'| \; , \tag{8.55}$$

i.e., (8.48) has weakness greater than 0, which is equivalent to

$$d_1 + d_2 - d - |N'| - 2|M'| - |Q'| - |R'| + 1 \leq |M| + |N| + |P| - 1 \; . \tag{8.56}$$

We sum up (8.56) and (8.54), and obtain $d \leq |M| - 1$. In that case we add again (8.46) to the sum (8.47), resp. (8.48), obtained by the input refutation following Algorithm 8.3.12, which yields

$$\begin{array}{c} \overline{Q} + \overline{R} + 2 \cdot \{\overline{N} + \overline{P} + \overline{T}\}^{|N|+|P|+|T|-1} \quad \geq \\ |N| + |P| + |Q| + |R| + 2|T| - 2d + d_1 + d_2 - |N'| - 2|M'| - |Q'| - |R'| \; . \end{array} \tag{8.57}$$

We have shown unsatisfiability if (8.57) is $\perp$, i.e., if

$$\begin{array}{c} |Q| + |R| + 2|N| + 2|P| + 2|T| - 2 \quad < \\ |N| + |P| + |Q| + |R| + 2|T| - 2d + d_1 + d_2 - |N'| - 2|M'| - |Q'| - |R'| \end{array} \tag{8.58}$$

holds. We assume the negation of (8.58), which simplifies to

$$2 \cdot d \geq d_1 + d_2 - |N| - |P| - |N'| - 2|M'| - |Q'| - |R'| + 2 \ . \tag{8.59}$$

Since (8.59) contradicts (8.54), we know that (8.57) is $\perp$.
Second, suppose that

$$|M| + |N| + |P| - 1 = d_1 + d_2 - d - |N'| - 2|M'| - |Q'| - |R'| \ , \tag{8.60}$$

thus we have

$$d_1 + d_2 - d - |N'| - 2|M'| - |Q'| - |R'| \leq |M| + |N| + |P| - 1 \ . \tag{8.61}$$

We sum up (8.61) and (8.54), and obtain $d \leq |M|$, which is equivalent to $d < |M| + 1$, which is (8.52). But then input refutation, following Algorithm 8.3.12, exhibits unsatisfiability. □

Example 8.3.14 We apply Algorithm 8.3.12 to the previous examples in this section.

[a] We again consider Example 8.3.9. We have $N' = \{X_5, X_6, X_7\}$, $R = \{X_1, X_2\}$, $Q = \{X_3, X_4\}$, $T = \{X_8\}$, and the other sets are empty. The first condition of Algorithm 8.3.12, i.e., (8.49), holds.

$$\begin{array}{rcl} |M| + |N| + |P| & \leq & d_1 + d_2 - d - |N'| - 2|M'| - |Q'| - |R'| \\ 0 + 0 + 0 & \leq & 2 + 3 - 2 - 3 - 0 - 0 - 0 \\ 0 & \leq & 0 \end{array}$$

[b] Consider again Example 8.3.10. We have $M = \{X_1, X_2, X_3\}$, $N' = \{X_8, X_9, X_{10}\}$, $Q = \{X_4, X_5\}$, $R = \{X_6\}$, $R' = \{X_7\}$, $T = \{X_{11}\}$, and the other sets are empty. The second condition of Algorithm 8.3.12, i.e., (8.50), which is $d < |M| + 1 (\equiv 3 < 3 + 1 = 4)$, and (8.52), hold.

$$\begin{array}{rcl} |M| + |N| + |P| - 1 & = & d_1 + d_2 - d - |N'| - 2|M'| - |Q'| - |R'| \\ 3 + 0 + 0 - 1 & = & 4 + 5 - 3 - 3 - 0 - 0 - 1 \\ 2 & = & 2 \end{array}$$

[c] Consider again Example 8.3.11. We have $M = \{X_1, X_2, X_3, X_4, X_5\}$, $N' = \{X_6\}$, $Q = \{X_7\}$, $R = \{X_8\}$, $T = \{X_9\}$, and the other sets are empty. The third condition of Algorithm 8.3.12, i.e.,

$$d \leq |M| - 1 (\equiv 3 \leq 5 - 1 = 4) \ ,$$

and (8.58) hold.

$$\begin{array}{rcl} 2d & < & d_1 + d_2 + |N| + |P| - |N'| - 2|M'| - |Q'| - |R'| + 2 \\ 6 & < & 3 + 3 + 0 + 0 - 1 - 0 - 0 - 0 + 2 \\ 6 & < & 7 \end{array}$$

Thus, Algorithm 8.3.12 finds a simplifying resolvent if there is one dominated by the sum of two extended clauses in S. The algorithm is of $O(|S|^2)$, since each pair of extended clauses in S needs to be considered. After splitting the set of literals in the extended clauses the test of whether there is a simplifying extended clause is reduced to simple arithmetic expressions only involving the cardinality of the split sets.

Restricted Two Step Input Refutation

Algorithm 8.3.12 still considers all possible pairs of extended clauses in S. The advantage of the presented two step input refutation is that we can start adding one extended clause and then skip cases that are not *promising*. Since the goal is to generate $\perp$ or force fixing literals, we may restrict ourselves to cases, where the number of literals of the left-hand side is already close to the right-hand side. We reformulate Algorithm 8.3.12 such that first one extended clause is added. The second extended clause is only added if a function `promising` decides to do so. Furthermore, we include the check for one step input resolution.

Algorithm 8.3.15 *Sequential One and Two Step Input Resolution*

seq_twostep_simpres($L \geq d, S$)
  forall $L' \geq d_1 \in S$ **do**
    $rhs_1 := |L| - d + d_1 - |L \cap L'|$
    $X, Y := \overline{L} \setminus (\overline{L \cap L'}), L' \setminus (L \cap L')$
    if $|X \cap Y| \geq 1 \wedge |X| + |Y| - 1 \leq rhs_1$ **then**
      (* one step input resolution *)
      $L_i \in \overline{X \cap Y}$
      return $L \setminus \{L_i\} \geq d$
    endif
    (* the current state is $X^{|X|-1} + Y \geq rhs_1$ *)
    if promising($L \geq d, L' \geq d_1, X, Y, rhs_1$) **then**
      forall $L'' \geq d_2 \in S$ **do**

(* add the second extended clause *)
$Y' := Y \setminus \overline{(\overline{Y} \cap L'')}$ (* resolve, no choice *)
$\hat{L}'' := L'' \setminus (\overline{Y} \cap L'')$ (* remaining from L'' *)
$X' := X \setminus \overline{(\overline{X} \cap \hat{L}'')}$ (* resolve, choice *)
$L''' := \hat{L}'' \setminus (\overline{X} \cap \hat{L}'')$ (* remaining from L'' *)
if $X' = \emptyset$ **then**
(* no simp. res. possible, since $N = P = T = \emptyset$ *)
else
$rhs_2 := rhs_1 + d_2 - |\overline{X} \cap \hat{L}''| - |(\overline{Y} \cap L''|$
$M := L \cap L' \cap L''$
(* check if $\perp$; current state $X'^{|X'|-1} + Y' + L''' \geq rhs$ *)
if $(|X'| + |Y'| + |L'''| \leq rhs_2) \vee$ (* (a) *)
$(|X'| + |Y'| + |L'''| - 1 = rhs_2 \wedge d < |M| + 1)$ (* (b) *)
then
$L_i \in L \cap ((L' \cap \overline{L''}) \cup (\overline{L'} \cap L'')) \vee L_i \in L \setminus (L' \cup L'')$
return $L \setminus \{L_i\} \geq d$
endif
if $d \leq |M| - 1$ **then** (* (c) *)
(* add $\overline{L}^{|L|-1} \geq |L| - d$ again, X', Y' not affected *)
$L'''' := L''' \setminus (L''' \cap L)$ (* resolve *)
$rhs_3 := rhs_2 + |L| - d - |L''' \cap L|$
if $|X'| + |Y'| + |L''''| + |\overline{L} \setminus \overline{L''' \cap L}| - 2 < rhs$ **then**
$L_i \in L \cap ((L' \cap \overline{L''}) \cup (\overline{L'} \cap L'')) \vee L_i \in L \setminus (L' \cup L'')$
return $L \setminus \{L_i\} \geq d$
endif
endif
endif
endforall
endif
endforall
return $\perp$
end `seq_twostep_simpres`

If the function `promising` always succeeds, i.e., all pairs of extended clauses are considered, then Algorithm 8.3.15 is equivalent to Algorithm 8.3.12. Note that:

$$\begin{aligned} L \cap L' &= M \cup N \cup Q \\ rhs_1 &= |P| + |R| + |T| - d + d_1 \\ X &= \overline{P} \cup \overline{R} \cup \overline{T} \end{aligned}$$

$$\begin{aligned}
Y &= M' \cup \overline{P} \cup N' \cup Q' \\
\overline{Y} \cap L'' &= P \cup \overline{N}' \\
Y' &= M' \cup Q' \\
\hat{L}'' &= M \cup M' \cup \overline{N} \cup R \cup R' \\
\overline{X} \cap \hat{L}'' &= R \\
X' &= \overline{P} \cup \overline{T} \\
L''' &= M \cup M' \cup \overline{N} \cup R' \\
rhs_2 &= |T| - d + d_1 + d_2 - |N'| \\
|X'| + |Y'| + |L'''| &= |P| + |T| + 2|M'| + |Q'| + |M| + |N| + |R'|
\end{aligned}$$

Therefore, we know that $|X'| + |Y' + |L'''| \leq rhs_2$ is equivalent to (8.49) and that $|X'| + |Y' + |L'''| - 1 = rhs_2$ is equivalent to (8.50). Furthermore,

$$\begin{aligned}
L''' \cap L &= M \ , \\
L'''' &= M' \cup \overline{N} \cup R' \ , \\
rhs_3 &= 2|T| - 2d + d_1 + d_2 - |N'| + |N| + |P| + |Q| + |R| \ , \\
\overline{L} \setminus \overline{L''' \cap L} &= N \cup P \cup Q \cup R \cup T \ ,
\end{aligned}$$

and

$$\begin{aligned}
|X'| + |Y'| + |L''''| + |\overline{L} \setminus \overline{L''' \cap L}| &= \\
2|P| + 2|T| + 2|M'| + |Q'| + 2|N| + |R'| + |Q| + |R| \ .
\end{aligned}$$

Thus, $|X'| + |Y'| + |L''''| + |\overline{L} \setminus \overline{L''' \cap L}| - 2 < rhs_3$ simplifies to

$$|P| + 2|M'| + |Q'| + |N| + |R'| - 2 < -2d + d_1 + d_2 - |N'| \ ,$$

which is equivalent to (8.58).

The main advantage of Algorithm 8.3.15 w.r.t. Algorithm 8.3.12 is that we can filter out combinations $\overline{L}^{|L|-1} + L' \geq |L| - d + d_1$ that are not likely to lead to the desired contradiction yielding a simplifying resolvent. The function `promising` decides whether a combination should be considered.

- Note that at least one of the extended clauses $L' \geq d_1$ and $L'' \geq d_2$ needs to contain a literal in L, since otherwise the sum $L' + L'' \geq d_1 + d_2$ cannot dominate $L \setminus \{L_i\} \geq d$, except if it is $\perp$. Since we consider each extended clause in S as a candidate for $L' \geq d_1$ as well as for $L'' \geq d_2$, we can reject all combinations $\overline{L}^{|L|-1} + L' \geq |L| - d + d_1$ with $L' \cap L = \emptyset$. Thus, `promising` should always include the test

if $L' \cap L = \emptyset$ **then** `return false` **endif**

and Algorithm 8.3.15 remains still equivalent to Algorithm 8.3.12.

- We want to simplify each extended clause in S. Since checking for a simplifying resolvent is $O(|S|^2)$ for each extended clause, we have $O(|S|^3)$ in the worst case when checking for each extended clause in S whether there is a simplifying resolvent for it. We propose to reduce the computational effort by rejecting some combinations that will probably not lead to a simplifying resolvent.

  A possible measure is $|X| + |Y| - 1 - rhs$, the difference between the right-hand side and the number of literals on the left-hand side. Since we obtain $\perp$, resp. fixing literals, if the number of literals is greater than, resp. equal to, the right-hand side, we can include

  if $|X| + |Y| - 1 - rhs > b$ **then** `return false` **endif**

  in `promising` for filtering out combinations where for more than b literals their negation has to be added after adding the first extended clause. The smaller b, the better the probability that we will find a contradiction and the more cases are filtered out. For example, with $b = 2$ all presented examples still work. Note that we may lose simplifying resolvents with that restriction. An improvement is to apply the filtering only if $d \geq |L \cap L'|$, since otherwise we may lose simplifying resolvents constructed by adding twice $\overline{L}^{|L|-1} \geq |L| - d$.

8.4 EQUATIONS BETWEEN LITERALS

An equation $L_i = L_j$ between two literals L_i, L_j is equivalent to the set of extended clauses

$$\{L_i + \overline{L_j} \geq 1, \overline{L_i} + L_j \geq 1\} \ . \tag{8.62}$$

Since the constraints we work with are only linear pseudo-Boolean inequalities and extended clauses, the best we can get whenever $L_i = L_j$ is dominated by the current constraint store, is that we have two different extended clauses in the constraint store, each of them dominating one of the extended clauses in (8.62). Note that for (8.62) no further simplifications apply.

But, the information $L_i = L_j$ can be used in order to simplify the constraint store. We can replace each occurrence of L_i by L_j and each occurrence of $\overline{L_i}$ by $\overline{L_j}$ without changing the solution space. After bringing the formulas in the constraint store into pseudo-Boolean normal form subsequent simplifications may be possible.

Example 8.4.1 Consider the linear pseudo-Boolean inequality

$$5 \cdot X_1 + 4 \cdot X_2 + 3 \cdot X_3 + 2 \cdot X_4 \geq 6$$

and suppose that $X_1 = X_2$ holds. After applying the equation to the linear pseudo-Boolean inequality and normalization we obtain

$$6 \cdot X_2 + 3 \cdot X_3 + 2 \cdot X_4 \geq 6 \ ,$$

which dominates $X_2 \geq 1$, i.e., we can fix X_2, and thus X_1, to 1. Given only the extended clauses $X_1 + \overline{X_2} \geq 1$ and $\overline{X_1} + X_2 \geq 1$ the same deduction is possible, but more steps are needed.

Transformation of $5 \cdot X_1 + 4 \cdot X_2 + 3 \cdot X_3 + 2 \cdot X_4 \geq 6$ gives

$$\{X_1 + X_2 \geq 1, X_1 + X_2 + X_3 + X_4 \geq 2\} \ .$$

Resolving $X_1 + \overline{X_2} \geq 1$ and $X_1 + X_2 \geq 1$ then generates $X_1 \geq 1$ and fixing X_1 to 1 in $\overline{X_1} + X_2 \geq 1$ then fixes X_2 to 1.

The crucial point in Example 8.4.1 is that we have to transform completely a linear pseudo-Boolean inequality and that we have to search for simplifying resolvents. Since transforming a linear pseudo-Boolean inequality is a computationally hard task, and since it is not sure that any simplifications or fixings of literals will happen, the task may be delayed. On the other hand, the application of an equality between Boolean variables is always useful, since the problem dimension is reduced, and thus should be performed immediately.

8.4.1 Equation Detection

Since an equation between literals simplifies the current constraint store, we show how to detect some equations of the form

$$L_i = L_j \tag{8.63}$$

dominated by a set of extended clauses. Since detecting equations is NP-complete[3], we restrict ourselves to the case that each of the two extended

[3] A problem is unsatisfiable if $X = \overline{X}$ can be derived.

clauses

$$L_i + \overline{L_j} \geq 1 \tag{8.64}$$
$$\overline{L_i} + L_j \geq 1 \tag{8.65}$$

are dominated by an extended clause in a set of extended clauses S. We thus overcome the loss of naturally expressing equations in our solved form.

Since both (8.64) and (8.65) have weakness 1, we know by Lemma 8.1.2 that if there are extended clauses in S dominating (8.64) and (8.65), then they must also have weakness 1[4]. Thus, it is sufficient to consider only the set of extended clauses, i.e., cliques (see Definition 7.7.5),

$$S' := \{L \geq |L| - 1 \mid L \geq |L| - 1 \in S\} .$$

Given a set of extended clauses S' containing only extended clauses with weakness 1, we show how to decide whether there are extended clauses (8.64) and (8.65), each of them dominated by one extended clause in S'.

An extended clause $L \geq |L| - 1$ dominates the classical clauses $L_i + L_j \geq 1$ for all $L_i, L_j \in L$ with $L_i \neq L_j$. Let us collect for all literals L_i the set of literals L_j such that $L_i + L_j \geq 1$. Suppose that these sets are stored in an array lits, indexed by the set of all literals whose variables occur in S'. Then we know that $L_i + \overline{L_j} \geq 1$ is dominated by an extended clause in S' if and only if $\overline{L_j} \in \texttt{lits}[L_i]$ and that $\overline{L_i} + L_j \geq 1$ is dominated by an extended clause in S' if and only if $L_j \in \texttt{lits}[\overline{L_i}]$. Thus, we can deduce $L_i = L_j$ if

$$L_j \in \texttt{lits}[\overline{L_i}] \cap \overline{\texttt{lits}[L_i]} . \tag{8.66}$$

Note that if $L_i = L_j$, then also $L_j = L_i$ and thus we can deduce $L_i = L_j$ if

$$L_i \in \texttt{lits}[\overline{L_j}] \cap \overline{\texttt{lits}[L_j]} .$$

We can save space and computation time by just trying to detect equations $L_i = L_j$, where $L_i \lhd L_j$ and condition (8.66) holds. The ordering $\lhd$ for literals was defined on page 164. Note that if $X_i \lhd X_j$, then $L_k \lhd L_l$ for all $L_k \in \{X_i, \overline{X_i}\}$ and for all $L_l \in \{X_j, \overline{X_j}\}$. Therefore, it is sufficient to store only literals L_j in $\texttt{lits}[L_i]$ with $L_i \lhd L_j$. Since it is convenient to look for equations $L_i = \overline{L_j}$, which is obviously no restriction, condition (8.66) becomes

$$L_j \in \overline{\texttt{lits}[\overline{L_i}]} \cap \texttt{lits}[L_i] . \tag{8.67}$$

[4]Note that extended clauses with weakness 0 fix all their literals. Thus, they are already removed and the fixed literals are collected elsewhere.

Next, we present an incremental algorithm that updates the array `lit` given a new clique inequality $L \geq |L| - 1$ and returns all equations between literals that are deducible.

Algorithm 8.4.2 *Equation Detection*

```
eq_detect(L ≥ |L| − 1)
        Eq := ∅
        forall L_i, L_j ∈ L ∧ L_i ◁ L_j do
                if L_j ∉ lit[L_i]then*
                        lit[L_i] := lit[L_i] ∪ {L_j}
                        if ¬L_j ∈ lits[¬L_i] then
                                Eq := Eq ∪ {L_i = ¬L_j}
                        endif
                endif
        endforall
        return Eq
end eq_detect
```

Algorithm 8.4.2 should be called with $\texttt{eq\_detect}(L \geq |L| - 1)$ whenever a new clique $L \geq |L| - 1$ enters the constraint store. The returned equations are immediately applied, i.e., for each $L_i = \overline{L_j} \in \texttt{eq\_detect}(L \geq |L| - 1)$ we replace in all linear pseudo-Boolean inequalities and extended clauses in the constraint store the literals L_i by $\overline{L_j}$ and the literals $\overline{L_i}$ by L_j. After bringing the updated constraints into normal form other simplifications may arise.

Note that after detecting an equation $L_i = \overline{L_j}$ we can immediately update $\texttt{lits}[L_i]$ by adding $\texttt{lits}[\overline{L_j}]$ and update $\texttt{lits}[\overline{L_i}]$ by adding $\texttt{lits}[L_j]$ and thus even more equations may immediately be detected. Since the returned equations are applied to the original constraint store, we get new cliques and the remaining equations are detected in the next round. In order to speed up searching and save space all literals L_i and $\overline{L_i}$ in all $\texttt{lits}[L_j]$ with $L_j \lhd L_i$ should be deleted.

8.4.2 Fixing Literals

Since we already maintain the data structure `lits`, we can use it also for the detection of fixed literals. Suppose there are literals L_i and L_j such that L_j and $\overline{L_j}$ are in $\texttt{lits}[L_i]$. Therefore, the extended clauses $L_i + \overline{L_j} \geq 1$ and

$L_i + L_j \geq 1$ are dominated by the current constraint store. Since resolving on the two extended clauses gives $L_i \geq 1$, and thus fixes L_i to 1, we have a simple method for checking whether there are literals that can be fixed. Similarly, we can deduce from $L_j \in \texttt{lits}[L_i]$ and $L_j \in \texttt{lits}[\overline{L_i}]$ that L_j needs to be fixed to 1.

We can insert in Algorithm 8.4.2 the lines

$$\textbf{if } \overline{L_j} \in \texttt{lits}[L_i] \textbf{ then } Eq := Eq \cup \{L_i = 1\} \textbf{ endif}$$
$$\textbf{if } L_j \in \texttt{lits}[\overline{L_i}] \textbf{ then } Eq := Eq \cup \{L_j = 1\} \textbf{ endif}$$

after the **then*** and thus additionally collect equations of the form $L_i = 1$ whenever deducible. Equations of that form are handled by pseudo-Boolean unit resolution, introduced in Chapter 5.

9

LINEARIZATION

In order to handle non-linear pseudo-Boolean constraints with a constraint solver for linear pseudo-Boolean inequalities, we have to linearize the non-linear pseudo-Boolean constraints in advance. It is sufficient to consider only non-linear pseudo-Boolean inequalities in normal form like in Definition 3.1.8, since others can be brought into that form with (3.3). Thus, the problem is to compute a set of linear pseudo-Boolean inequalities equivalent to a non-linear pseudo-Boolean inequality in normal form. After a review of well-known linearization techniques, we present a linearization method that fits well into the context of constraint logic programming.

9.1 PREVIOUS WORK

The first linearization method for non-linear pseudo-Boolean constraints was introduced by Fortet [For59, For60] and Dantzig [Dan60]. Consider a non-linear pseudo-Boolean inequality in normal form

$$\sum_{I \in \Omega} c_I \prod_{i \in I} X_i \geq d \tag{9.1}$$

like in Definition 3.1.8. The idea of Fortet and Dantzig is to replace each product term $\prod_{i \in I} X_i$ by a new Boolean variable X_I yielding

$$\sum_{I \in \Omega} c_I X_I \geq d \tag{9.2}$$

and add the linear pseudo-Boolean inequalities

$$X_I + \sum_{i \in I} \overline{X_i} \geq 1 \tag{9.3}$$

$$|I| \cdot \overline{X_I} + \sum_{i \in I} X_i \geq |I| \tag{9.4}$$

for all $I \in \Omega$. Since (9.3) forces $X_I = 1$ if $X_i = 1$ for all $i \in I$, and (9.4) forces $X_I = 0$ if $X_i = 0$ for some $i \in I$, we know that the non-linear pseudo-Boolean inequality (9.1) is equivalent to the set of linear pseudo-Boolean inequalities

$$\{\sum_{I \in \Omega} c_I X_I \geq d\} \cup \{|I| \cdot \overline{X_I} + \sum_{i \in I} X_i \geq |I|, X_I + \sum_{i \in I} \overline{X_i} \geq 1 \mid \text{ for all } I \in \Omega\} \ . \tag{9.5}$$

Therefore, we have a simple method for replacing a non-linear pseudo-Boolean inequality by an equivalent set of linear pseudo-Boolean inequalities. Glover and Wolsey [GW73] modify the method in order to decrease the number of linear pseudo-Boolean inequalities in (9.5) while preserving equivalence.

The introduction of new Boolean variables is the main problem of the Fortet-Dantzig linearization, since the dimension of the problem is increased. As reported by Hansen et al. [HJM93] the Fortet-Dantzig linearization is adequate if there are only a few non-linear pseudo-Boolean terms. In the context of constraint logic programming the Fortet-Dantzig linearization is not suitable, since the new Boolean variables have to be eliminated at the end of computation.

Balas and Mazzola [BM84a] present linearization techniques based on the work of Granot and Hammer [GH75] that do not introduce new Boolean variables. Their basic idea is to generate sufficiently many cutting plane inequalities until equivalence is established [HJM93, BM84a, BM84b]. We develop a linearization method that takes into account the current constraint store in order to obtain immediately stronger linear pseudo-Boolean inequalities whenever possible.

9.2 A LINEARIZATION METHOD

We present a linearization method that fits well into the context of constraint logic programming, since the current constraint store can be used to influence the linearization process. Furthermore, the linearization method does not introduce new Boolean variables. The principal idea is to successively split a non-linear pseudo-Boolean inequality into two others that are in some sense *more linear.* The process is repeated until only linear pseudo-Boolean inequalities remain. While splitting the non-linear pseudo-Boolean inequalities, we need to derive lower bounds for non-linear pseudo-Boolean terms. By taking into account the constraint store when deriving these bounds, we can derive stronger linear pseudo-Boolean inequalities. Furthermore, the method coincides nicely with the compact representation of linear pseudo-Boolean constraints presented in Section 7.4.

We do not require that the non-linear pseudo-Boolean inequality is in normal form, but we allow non-linear pseudo-Boolean inequalities of the form

$$\sum_{I \in \Omega} c_I \prod_{i \in I} L_i \geq d \ , \tag{9.6}$$

where for all $i, j \in I$ we have $\mathsf{Var}(L_i) \neq \mathsf{Var}(L_j)$ for all $I \in \Omega$. The main difference w.r.t. the normal form of Definition 3.1.8 is that we allow literals and not only variables in the product terms. First, we show how to split the left-hand side of (9.6) w.r.t. a variable X_j occurring in (9.6).

Lemma 9.2.1 *A non-linear pseudo-Boolean inequality $\sum_{I \in \Omega} c_I \prod_{i \in I} L_i \geq d$ containing a variable X_j is equivalent to the inequality*

$$X_j \cdot \Big(\sum_{I \in \Omega : \forall i \in I : L_i \neq \overline{X_j}} c_I \prod_{i \in I : L_i \neq X_j} L_i \Big) + \overline{X_j} \cdot \Big(\sum_{I \in \Omega : \forall i \in I : L_i \neq X_j} c_I \prod_{i \in I : L_i \neq \overline{X_j}} L_i \Big) \geq d \ . \tag{9.7}$$

Proof. The pseudo-Boolean term that is multiplied by X_j is the sum over all product terms of (9.6), where either X_j occurs or neither X_j nor $\overline{X_j}$ occurs. Similarly, the pseudo-Boolean term that is multiplied by $\overline{X_j}$ is the sum over all product terms of (9.6), where either $\overline{X_j}$ occurs or neither X_j nor $\overline{X_j}$ occurs. In both terms X_j and $\overline{X_j}$ is deleted. By multiplying X_j, resp. $\overline{X_j}$, into the corresponding pseudo-Boolean term we obtain all product terms of the left-hand side of (9.6), where X_j or $\overline{X_j}$ occurs, plus the pseudo-Boolean terms

$$X_j \cdot c_I \prod_{i \in I} L_i + \overline{X_j} \cdot c_I \prod_{i \in I} L_i \tag{9.8}$$

for all product terms of the left-hand side of (9.6) not containing X_j or $\overline{X_j}$. Since (9.8) is equivalent to $c_I \prod_{i \in I} L_i$, we have shown equivalence of (9.6) and (9.7). □

Theorem 9.2.2 *A non-linear pseudo-Boolean inequality* $\sum_{I\in\Omega} c_I \prod_{i\in I} L_i \geq d$ *is equivalent to the two pseudo-Boolean inequalities*

$$(d - l_1) \cdot \overline{X_j} + \sum_{I\in\Omega:\forall i\in I:L_i\neq\overline{X_j}} c_I \prod_{i\in I:L_i\neq X_j} L_i \geq d \tag{9.9}$$

$$(d - l_2) \cdot X_j + \sum_{I\in\Omega:\forall i\in I:L_i\neq X_j} c_I \prod_{i\in I:L_i\neq\overline{X_j}} L_i \geq d \ , \tag{9.10}$$

where l_1 *is a lower bound for*

$$\sum_{I\in\Omega:\forall i\in I:L_i\neq\overline{X_j}} c_I \prod_{i\in I:L_i\neq X_j} L_i \quad \textit{subject to} \quad \{(9.6), X_j = 0\}$$

and l_2 *is a lower bound for*

$$\sum_{I\in\Omega:\forall i\in I:L_i\neq X_j} c_I \prod_{i\in I:L_i\neq\overline{X_j}} L_i \quad \textit{subject to} \quad \{(9.6), X_j = 1\} \ .$$

Proof. We abbreviate (9.7) by

$$X_j \cdot t_1 + \overline{X_j} \cdot t_2 \geq d \ .$$

Thus, inequality (9.9) is equivalent to $(d - l_1) \cdot \overline{X_j} + t_1 \geq d$ and (9.10) is equivalent to $(d - l_2) \cdot X_j + t_2 \geq d$, where l_1 is a lower bound for t_1 subject to $\{(9.6), X_j = 0\}$ and l_2 is a lower bound for t_2 subject to $\{(9.6), X_j = 1\}$. Note that if the constraint $\{(9.6), X_j = c\}$ is unsatisfiable for $c = 0$, resp. $c = 1$, then l_1, resp. l_2, is arbitrary. We show

$$\mathsf{Ext}(X_j \cdot t_1 + \overline{X_j} \cdot t_2 \geq d) = \mathsf{Ext}(\{(d - l_1) \cdot \overline{X_j} + t_1 \geq d, (d - l_2) \cdot X_j + t_2 \geq d\})$$

in order to prove the theorem.

$\subseteq$: Let α be in $\mathsf{Ext}(X_j \cdot t_1 + \overline{X_j} \cdot t_2 \geq d)$. First, suppose that $\alpha(X_j) = 1$. Then $\alpha(X_j \cdot t_1 + \overline{X_j} \cdot t_2 \geq d)$ simplifies to $\alpha(t_1) \geq d$. Since $\alpha((d - l_1) \cdot \overline{X_j} + t_1) = \alpha(t_1)$ and $\alpha(t_1) \geq d$, we have $\alpha \in \mathsf{Ext}((d - l_1) \cdot \overline{X_j} + t_1 \geq d)$. Furthermore, $\alpha((d - l_2) \cdot X_j + t_2) \geq d$ simplifies to $d - l_2 + \alpha(t_2) \geq d$ and after bringing the constants to the right hand side we have $\alpha(t_2) \geq l_2$. Since l_2 is a lower bound for t_2 for all $\alpha' \in \mathsf{Ext}(X_j \cdot t_1 + \overline{X_j} \cdot t_2 \geq d)$ with $\alpha'(X_j) = 1$, we know that $\alpha(t_2) \geq l_2$. Thus, $\alpha \in \mathsf{Ext}((d - l_2) \cdot X_j + t_2 \geq d)$. Similarly, for $\alpha(X_j) = 0$.

$\supseteq$: Let α be in $\mathsf{Ext}(\{(d - l_1) \cdot \overline{X_j} + t_1 \geq d, (d - l_2) \cdot X_j + t_2 \geq d\})$. First, suppose that $\alpha(X_j) = 1$. From $\alpha((d - l_1) \cdot \overline{X_j} + t_1) \geq d$ we derive $\alpha(t_1) \geq d$ and thus $\alpha(X_j \cdot t_1 + \overline{X_j} \cdot t_2) \geq d$, because $\alpha(\overline{X_j} \cdot t_2) = 0$. Therefore, $\alpha \in \mathsf{Ext}(X_j \cdot t_1 + \overline{X_j} \cdot t_2 \geq d)$. Similarly, for $\alpha(X_j) = 0$. □

Theorem 9.2.2 enables us to split a non-linear pseudo-Boolean inequality into two others. In order to give a complete linearization algorithm, we define a measure of non-linearity of non-linear pseudo-Boolean inequalities.

Definition 9.2.3 A Boolean variable X_j has a *non-linear* occurrence in a non-linear pseudo-Boolean inequality $\sum_{I\in\Omega} c_I \prod_{i\in I} L_i \geq d$ if there is an $I \in \Omega$ such that $|I| \geq 2$ and there exists an $i \in I$ such that $\mathsf{Var}(L_i) = X_j$. A Boolean variable has a *linear* occurrence in a non-linear pseudo-Boolean inequality if it occurs in the inequality, but has no non-linear occurrence. We call $\sum_{I\in\Omega} \max(0, |I| - 1)$ the *non-linearity* of a non-linear pseudo-Boolean inequality $\sum_{I\in\Omega} c_I \prod_{i\in I} L_i \geq d$.

Since t_1 and t_2 of Theorem 9.2.2 do not contain the variable X_j, the occurrence of the variable X_j in the pseudo-Boolean inequalities (9.9) and (9.10) is linear. If we have split on a variable X_j having a non-linear occurrence in (9.6), then the resulting two pseudo-Boolean inequalities (9.9) and (9.10) are *more linear* than (9.6), since the non-linearity of (9.9) and (9.10) is smaller than the non-linearity of (9.6). Note that linear product terms, i.e., product terms containing at most one literal, do not count for the non-linearity. Furthermore, for each non-linear occurrence of X_j in a product term, the number of literals in this product term is decreased. If there are no non-linear occurrences of Boolean variables, then the pseudo-Boolean inequality is linear. We obtain the following linearization algorithm transforming a non-linear pseudo-Boolean inequality into an equivalent set of linear pseudo-Boolean inequalities.

Algorithm 9.2.4 *Linearization of non-linear pseudo-Boolean inequalities*

```
linearize(t ≥ d)
    if ∃X_j having a non-linear occurrence in t ≥ d then
        reformulate t ≥ d to X_j · t_1 + ¬X_j · t_2 ≥ d (* Lemma 9.2.1 *)
        l_1 := nlpbt_lower_bound(t_1, t ≥ d, {¬X_j})
        l_2 := nlpbt_lower_bound(t_2, t ≥ d, {X_j})
        return (linearize((d − l_1) · ¬X_j + t_1 ≥ d) ∪
                linearize((d − l_2) · X_j + t_2 ≥ d))
    else
        return {t ≥ d}
    endif
end linearize
```

We assume the existence of a function nlpbt_lower_bound($t', t \geq d, F$) that computes a lower bound for a non-linear pseudo-Boolean term t' subject to the set of pseudo-Boolean constraints $\{t \geq d, F \geq |F|\}$, where F is a set of literals to be fixed. A naive implementation of `nlpbt_lower_bound` just computes the sum of all negative coefficients of t'. Better lower bounds, and thus stronger linear pseudo-Boolean inequalities, are obtained by taking into account the constraints $t \geq d$ and $F \geq |F|$.

Example 9.2.5 Let

$$A \cdot B + A \cdot C + B \cdot C \geq 1 \tag{9.11}$$

be the non-linear pseudo-Boolean inequality that is to be linearized. First, we split on A and obtain

$$A \cdot (B + C + B \cdot C) + \overline{A} \cdot (B \cdot C) \geq 1 \ .$$

A naive lower bound for both terms is 0. Thus, by Theorem 9.2.2, we obtain the set of pseudo-Boolean inequalities

$$\{\overline{A} + B + C + B \cdot C \geq 1, A + B \cdot C \geq 1\} \ ,$$

which is equivalent to $A \cdot B + A \cdot C + B \cdot C \geq 1$. Next, we split the first non-linear pseudo-Boolean inequality on B and obtain

$$B \cdot (\overline{A} + 1 + 2 \cdot C) + \overline{B} \cdot (\overline{A} + C) \geq 1$$

With the lower bound 1 for the first and 0 for the second term we get

$$\{(1-1) \cdot \overline{B} + \overline{A} + 1 + 2 \cdot C \geq 1, B + \overline{A} + C \geq 1\} \ ,$$

which is equivalent to $\overline{A} + B + C + B \cdot C \geq 1$. Note that the first inequality is $\top$ and can be eliminated. It remains to linearize $A + B \cdot C \geq 1$. We split on B and obtain

$$B \cdot (A + C) + \overline{B} \cdot A \geq 1 \ .$$

According to Theorem 9.2.2 we derive with the lower bound 0 the set of pseudo-Boolean inequalities

$$\{\overline{B} + A + C \geq 1, B + A \geq 1\} \ ,$$

which is equivalent to $A + B \cdot C \geq 1$. There are no more pseudo-Boolean inequalities that need to be linearized. Therefore,

$$\{\overline{A} + B + C \geq 1, A + \overline{B} + C \geq 1, A + B \geq 1\}$$

is equivalent to $A \cdot B + A \cdot C + B \cdot C \geq 1$.

Let us again linearize (9.11), but take into account the constraints while calculating the lower bounds. We again split $A \cdot B + A \cdot C + B \cdot C \geq 1$ on A and obtain

$$A \cdot (B + C + B \cdot C) + \overline{A} \cdot (B \cdot C) \geq 1 \ .$$

While calculating a lower bound for $B + C + B \cdot C$ we have the additional constraint

$$\{A \cdot B + A \cdot C + B \cdot C \geq 1, A = 0\} \ ,$$

which simplifies to $B \cdot C \geq 1$ and thus $B = C = 1$. Therefore, a lower bound for $B + C + B \cdot C$ is 3. Likewise, we calculate the lower bound 0 for $B \cdot C$ subject to $\{A \cdot B + A \cdot C + B \cdot C \geq 1, A = 1\}$, which simplifies to $B + C + B \cdot C \geq 1$. We obtain the set of pseudo-Boolean inequalities

$$\{2 \cdot A + B + C + B \cdot C \geq 3, A + B \cdot C \geq 1\} \ ,$$

which is equivalent to $A \cdot B + A \cdot C + B \cdot C \geq 1$. Next, we linearize $2 \cdot A + B + C + B \cdot C \geq 3$. We split on B and obtain

$$B \cdot (2 \cdot A + 1 + 2 \cdot C) + \overline{B} \cdot (2 \cdot A + C) \geq 3 \ .$$

We get the lower bound 5 for $2 \cdot A + 1 + 2 \cdot C$ subject to $2 \cdot A + C \geq 3$ and the lower bound 0 for $2 \cdot A + C$ subject to $2 \cdot B \geq 0$. According to Theorem 9.2.2 we know that

$$\{2 \cdot B + 2 \cdot A + 2 \cdot C \geq 4, 3 \cdot B + 2 \cdot A + C \geq 3\}$$

is equivalent to $2 \cdot A + B + C + B \cdot C \geq 3$. If we transform $2 \cdot B + 2 \cdot A + 2 \cdot C \geq 4$ with the method described in Chapter 7, we obtain the single extended clause $A + B + C \geq 2$. Transforming all other linear pseudo-Boolean inequalities obtained by the linearization process yields only extended clauses that are dominated by $A + B + C \geq 2$. Thus, the non-linear pseudo-Boolean inequality

$$A \cdot B + A \cdot C + B \cdot C \geq 1$$

is equivalent to the single extended clause

$$A + B + C \geq 2 \ .$$

9.3 DERIVING LOWER BOUNDS

Example 9.2.5 indicates that calculating good lower bounds is worthwhile. In principle we can calculate the best lower bound for t' subject to $t \geq d$ by solving the non-linear pseudo-Boolean optimization problem

$$\min t' \quad \text{subject to} \quad t \geq d \ . \tag{9.12}$$

Since this is computationally too expensive, we approximate the best possible lower bound as follows. First, we fix all literals that can be fixed. Afterwards, we check whether a product term has to evaluate to 1, i.e., if further fixings are implied. This process is repeated until no more obvious fixings apply. A lower bound is then obtained by summing up all remaining negative coefficients. Fixing literals in a non-linear pseudo-Boolean term is done by `nl_fixing` defined as:

$$\texttt{nl\_fixing}(\textstyle\sum_{I\in\Omega} c_I \prod_{i\in I} L_i \geq d, F) := \sum_{I'\in\Omega'} c_{I'} \prod_{i\in I'} L_i \geq d \quad \text{where} \\ \Omega' = \{I' \mid I \in \Omega \wedge \forall i \in I : \overline{L_i} \notin F \wedge I' = I \setminus F\} \tag{9.13}$$

If any simplifications are possible, like bringing constants on the right-hand side and grouping identical product terms, these are performed by bringing the non-linear pseudo-Boolean inequality into normal form. We relax the detection of fixed literals by viewing each product term as a single Boolean variable. Thus, the task becomes computationally manageable and is similar to `fixed` in (5.7).

$$\texttt{nl\_fixed}(\sum_{I\in\Omega} c_I \prod_{i\in I} L_i \geq d) := \begin{cases} \{L_j \mid \forall j \in J\} & \text{if } \sum_{I\in\Omega'\setminus\{J\}} c_I - c_J < d \\ & \text{where} \\ & \Omega' = \{I \mid I \in \Omega \wedge c_I \geq 0\}, \\ & J \in \Omega, \text{and} \\ & c_J \geq c_I \text{ for all } I \in \Omega \\ \bot & \text{otherwise} \end{cases} \tag{9.14}$$

Note that we can only consider positive coefficients. Hence, some obvious fixings may not be detected. We can overcome that problem by reformulating the non-linear pseudo-Boolean inequality such that only positive coefficients occur with help of the following equation.

$$c_I \cdot \prod_{i\in I} L_i = c_I \cdot \prod_{i\in I\setminus\{j\}} L_i + (-c_I) \cdot \overline{L_j} \cdot \prod_{i\in I\setminus\{j\}} L_i \quad \text{where } j \in I \tag{9.15}$$

Let c_I be negative and let us apply (9.15) recursively to the product term $c_I \cdot \prod_{i\in I\setminus\{j\}} L_i$. Since this product term contains only $|I| - 1$ literals, we end up with the negative constant c_I, which can be brought to the right-hand side, and a sum of product terms having the positive coefficient $-c_I$. Note that a product $c_I \cdot \prod_{i\in I} L_i$ with $c_I < 0$ expands to at most $|I|$ products with positive coefficients. Therefore, we have a method for reformulating a non-linear pseudo-Boolean inequality such that only positive coefficients occur and more fixings of literals will be detected in `nl_fixed`. Next, we give a possible implementation of `nlpbt_lower_bound`, which computes a lower bound for a non-linear pseudo-Boolean term t by first doing all possible fixings and then computing a naive lower bound.

Algorithm 9.3.1 *Lower bound for a non-linear pseudo-Boolean term*

```
nlpbt_lower_bound(t', t ≥ d, F)
    while F ≠⊥
        t ≥ d := nl_fixing(t ≥ d, F)
        t' := nlpbt_fixing(t', F)
        F := nl_fixed(t ≥ d)
    endwhile
    return naive_lower_bound(t', t ≥ d)
end nlpbt_lower_bound
```

We assume the existence of a function `nlpbt_fixing`, which is similar to `nl_fixing`, but fixes the literals in a pseudo-Boolean term only and collects the constants separately.

$$\begin{array}{l}\texttt{nlpbt\_fixing}(\sum_{I\in\Omega} c_I \prod_{i\in I} L_i, F) := \sum_{I'\in\Omega'} c_{I'} \prod_{i\in I'} L_i \quad \text{where} \\ \Omega' = \{I' \mid I \in \Omega \wedge \forall i \in I : \overline{L_i} \notin F \wedge I' = I \setminus F\}\end{array} \tag{9.16}$$

We also assume the existence of a function `naive_lower_bound` that returns a lower bound for a non-linear pseudo-Boolean term t'. A simple implementation of `naive_lower_bound` calculates the sum of all negative coefficients of t' plus the constant terms of t'.

Example 9.3.2 Consider again Example 9.2.5 and let us calculate a lower bound for $B + C + B \cdot C$ subject to $\{A \cdot B + A \cdot C + B \cdot C \geq 1, A = 0\}$ with `nlpbt_lower_bound`. We call

$$\texttt{nlpbt\_lower\_bound}(B + C + B \cdot C, A \cdot B + A \cdot C + B \cdot C \geq 1, \{\overline{A}\})$$

and get in the first iteration of the **while**-loop

$$\begin{array}{rcl} t \geq d & = & B \cdot C \geq 1, \\ t' & = & B + C + B \cdot C, \\ F & = & \{B, C\} \ . \end{array}$$

Since $F \neq \perp$, we continue and get in the second iteration of the **while**-loop

$$\begin{array}{rcl} t \geq d & = & \top, \\ t' & = & 3, \\ F & = & \perp \ . \end{array}$$

Thus, we return the naive lower bound 3. The presented `nlpbt_lower_bound` calculates the same lower bounds as in Example 9.2.5.

Next, we improve **naive_lower_bound** based on the observation that the non-linear pseudo-Boolean optimization problem (9.12) has a special form. Typically, many product terms that occur in t' also occur in t.

Example 9.3.3 Let

$$5 \cdot L_1 + 4 \cdot L_2 + 3 \cdot L_3 \cdot \overline{L_5} + 2 \cdot L_4 + L_5 \cdot L_6 \geq 6$$

be the non-linear pseudo-Boolean inequality to be linearized. We split on L_5 and obtain an optimal lower bound for $t' = 5 \cdot L_1 + 4 \cdot L_2 + 2 \cdot L_4 + L_6$ by solving the non-linear pseudo-Boolean optimization problem

$$\min 5 \cdot L_1 + 4 \cdot L_2 + 2 \cdot L_4 + L_6 \quad \text{subject to} \quad 5 \cdot L_1 + 4 \cdot L_2 + 3 \cdot L_3 + 2 \cdot L_4 \geq 6 \ .$$

We eliminate the product terms in the constraint that do not occur in the objective function, which yields the valid non-linear pseudo-Boolean inequality

$$5 \cdot L_1 + 4 \cdot L_2 + 2 \cdot L_4 \geq 3 \ ,$$

whose left-hand side is a subterm of the objective function. Thus, we can derive the lower bound 3 and get the linear pseudo-Boolean inequality

$$3 \cdot \overline{L_5} + 5 \cdot L_1 + 4 \cdot L_2 + 2 \cdot L_4 + L_6 \geq 6 \ .$$

Just summing up the constant and all negative coefficients gives the lower bound 0. Hence, with **naive_lower_bound** only the weaker linear pseudo-Boolean inequality

$$6 \cdot \overline{L_5} + 5 \cdot L_1 + 4 \cdot L_2 + 2 \cdot L_4 + L_6 \geq 6$$

can be constructed.

Many product terms are shared between the objective function and the left-hand side of the constraint, since the objective function is obtained by selecting all product terms not containing the negation of a specific literal L, whereas the left-hand side of the constraint is obtained by fixing L to 0. By taking into account these shared products terms, we obtain a valid lower bound for a pseudo-Boolean term

$$\sum_{I \in \Omega_1} c_I \prod_{i \in I} L_i + \sum_{I \in \Omega_2} c_I \prod_{i \in I} L_i + \sum_{I \in \Omega_3} c_I \prod_{i \in I} L_i$$

w.r.t.

$$\sum_{I \in \Omega_1} c_I \prod_{i \in I} L_i + \sum_{I \in \Omega_4} c_I \prod_{i \in I} L_i + \sum_{I \in \Omega_5} c_I \prod_{i \in I} L_i \geq d \ ,$$

where $c_I > 0$ for all $I \in \Omega_2 \cup \Omega_4$ and $c_I < 0$ for all $I \in \Omega_3 \cup \Omega_5$, is

$$d - \sum_{I \in \Omega_4} c_I + \sum_{I \in \Omega_3} c_I \ . \tag{9.17}$$

Note that Ω_1 represent all product terms that occur in both the objective function and the constraint. The set Ω_2, resp. Ω_3, represents the product term with positive coefficients, resp. negative coefficients in the objective function that do not occur in the constraint. Vice versa for the Ω_4 and Ω_5. Since the objective function and the left-hand side of the constraint typically share many product terms, we know that Ω_4 and Ω_3 are small and we may obtain a good lower bound. Thus, we propose to replace `naive_lower_bound` by `improved_lower_bound`, where we view a sum of product terms as a set of product terms.

Algorithm 9.3.4 *Improved Lower Bound*

> **improved_lower_bound**$(t', t \geq d)$
> $O_4 := \{c_I \mid c_I > 0 \wedge c_I \prod_{i \in I} L_i \in (t \setminus t')\}$
> $O_3 := \{c_I \mid c_I < 0 \wedge c_I \prod_{i \in I} L_i \in (t' \setminus t)\}$
> (* equation (9.17) *)
> $NLB_1 := d - \sum_{c_I \in O_4} c_I + \sum_{I \in O_3} c_I$
> (* sum of constants and negative coefficients *)
> $NLB_2 := \sum_{I: c_I \prod_{i \in I} L_i \in t' \wedge (c_I < 0 \vee I = \emptyset)} c_I$
> **return** $\max(NLB_1, NLB_2)$
> **end improved_lower_bound**

Example 9.3.5 Let

$$5 \cdot A + 4 \cdot B + 3 \cdot C + 2 \cdot D + E \cdot F \geq 10$$

be the non-linear pseudo-Boolean inequality to be linearized. Using the improved lower bounds of Algorithm 9.3.4 we get the set of linear pseudo-Boolean inequalities

$$\{5 \cdot A + 4 \cdot B + 3 \cdot C + 2 \cdot D + E \geq 10,$$
$$5 \cdot A + 4 \cdot B + 3 \cdot C + 2 \cdot D + F \geq 10\} \ .$$

Just summing up the constants and the negative coefficients yields the weaker linear pseudo-Boolean inequalities

$$\{10 \cdot \overline{E} + 5 \cdot A + 4 \cdot B + 3 \cdot C + 2 \cdot D + F \geq 10,$$
$$10 \cdot E + 5 \cdot A + 4 \cdot B + 3 \cdot C + 2 \cdot D \geq 10\} \ .$$

So, the better the lower bound, the stronger the linear pseudo-Boolean inequalities that result. By taking into account the constraint store when calculating the lower bounds, we perform *constrained linearization*, which is suitable in the context of constraint logic programming. For example, the set of fixed literals calculated in Algorithm 9.3.1 can be temporarily fixed in the current constraint store. Fixings that result from this operation then can be propagated into `nlpbt_lower_bound`, i.e., further literals can be fixed. Thus, stronger lower bounds w.r.t. the current constraint store result and stronger linear pseudo-Boolean inequalities are generated while equivalence w.r.t. the current constraint is still preserved.

The presented linearization algorithm compares well with the linearization method of Balas and Mazzola [BM84a, BM84b] on the examples found in their paper. They reformulate the non-linear pseudo-Boolean inequality (Example 6 [BM84a])

$$10 \cdot X_1 \cdot X_9 + 9 \cdot X_2 \cdot X_8 + 8 \cdot X_3 \cdot X_7 + 8 \cdot X_1 \cdot X_6 + 8 \cdot X_3 \cdot X_4 + 5 \cdot X_2 \cdot X_5 \leq 20$$

into an equivalent set of 20 cover inequalities. With our approach, not including improvements, we generate 4 linear pseudo-Boolean inequalities, that, when transformed, are equivalent to 25 extended clauses. They reformulate the non-linear pseudo-Boolean inequality (Example 3 [BM84b])

$$6 \cdot X_1 \cdot X_2 \cdot X_3 \cdot X_4 + 4 \cdot X_1 \cdot X_5 - 3 \cdot X_3 \cdot X_4 \cdot X_6 + 2 \cdot X_1 \cdot X_2 \cdot X_4 \cdot X_7 + 2 \cdot X_3 \cdot X_4 \cdot X_8 \leq 7$$

with an improved method into the single equivalent linear pseudo-Boolean inequality

$$7 \cdot \overline{X_1} + 7 \cdot \overline{X_2} + 7 \cdot \overline{X_3} + 7 \cdot \overline{X_4} + 4 \cdot \overline{X_5} + 3 \cdot \overline{X_6} + 2 \cdot \overline{X_7} + 2 \cdot \overline{X_8} \geq 7 \ .$$

Our method computes the same inequality.

9.4 FURTHER IMPROVEMENTS

The variable X_j on which to split in Theorem 9.2.2 is important for the efficiency of the presented algorithm. We propose to split on the variable having the largest number of non-linear occurrences, since then the sum of the non-linearity of the resulting two pseudo-Boolean inequalities is minimal.

The worst case occurs if each Boolean variable has exactly one occurrence in the non-linear pseudo-Boolean inequality to linearize. Then each Boolean variable, except one in each product term, has to be eliminated. Exactly for this case, we know of a better representation; the compact representation of linear pseudo-Boolean inequalities introduced in Section 7.4. Since we may have negative coefficients in a non-linear pseudo-Boolean inequality, we need to drop the condition of Definition 7.4.1 that the coefficients are all positive and ordered. Alternatively, we can reformulate a product term with a negative coefficient into one having only positive coefficients with help of (9.15).

Proposition 9.4.1 *Let the non-linear pseudo-Boolean inequality*

$$\sum_{I \in \Omega} c_I \prod_{i \in I} L_i \geq d \tag{9.18}$$

contain each Boolean variable only once, i.e., for all $i, j \in I$ we have $\mathsf{Var}(L_i) \neq \mathsf{Var}(L_j)$ for all $I \in \Omega$. W.l.o.g. assume that $c_I \geq 1$ for all $I \in \Omega$. Then (9.18) is equivalent to the compact pseudo-Boolean inequality set

$$\sum_{I \in \Omega} c_I \{L_i \mid i \in I\}^1 \geq d \ . \tag{9.19}$$

Proof. We show that the extension of (9.18) is the same as the extension of (9.19).

"$\subseteq$:" Let α be in the extension of (9.18). Let Ω_1 be those $I \in \Omega$ for which $\alpha(\prod_{i \in I} L_i) = 1$ and let Ω_2 be $\Omega \backslash \Omega_1$. Then $\sum_{I \in \Omega_1} c_I \geq d$ and for all $I \in \Omega_1$ we have $\alpha(L_i) = 1$ for all $i \in I$. Each linear pseudo-Boolean inequality in the expansion of (9.19) is of the form $\sum_{I \in \Omega_1} c_I L_{I_i} + \sum_{I \in \Omega_2} c_I L_{I_i} \geq d$ with $I_i \in I$ for all $I \in \Omega$. Since for all $I \in \Omega_1$ we have $\alpha(L_{I_i}) = 1$ for all $I_i \in I$, we know that $\alpha(\sum_{I \in \Omega_1} c_I L_{I_i}) = \sum_{I \in \Omega_1} c_I \geq d$, and thus $\alpha(\sum_{I \in \Omega_1} c_I L_{I_i} + \sum_{I \in \Omega_2} c_I L_{I_i}) \geq d$. Therefore, α is in the extension of each linear pseudo-Boolean inequality in the expansion of (9.19) and thus in the extension of (9.19).

"$\supseteq$:" We show that if α is not in the extension of (9.18), then α is not in the extension of (9.19). Let α be not in the extension of (9.18). Let Ω_1 be those $I \in \Omega$ for which $\alpha(\prod_{i \in I} L_i) = 0$ and let Ω_2 be $\Omega \setminus \Omega_1$ Then $\sum_{I \in \Omega_2} c_I < d$ and for each $I \in \Omega_1$ there is a literal $L_{I_i}, I_i \in I$, such that $\alpha(L_{I_i}) = 0$. Let us consider the linear pseudo-Boolean inequality in the expansion of (9.19), where we have selected exactly that literal L_{I_i} for each $I \in \Omega_1$ and an arbitrary literal for each $I \in \Omega_2$. Since then $\alpha(\sum_{I \in \Omega_1} c_I L_{I_i} + \sum_{I \in \Omega_2} c_I L_{I_i}) = \alpha(\sum_{I \in \Omega_2} c_I L_{I_i}) < d$, because $\alpha(\sum_{I \in \Omega_1} c_I L_{I_i}) = 0$, we know that α is not in the extension of that linear pseudo-Boolean inequality in the expansion of (9.19), and thus not in the extension of (9.19). □

We can use Proposition 9.4.1 in order to enhance our linearization method. We propose to split only on a Boolean variable if it has at least two occurrences and at least one of them is non-linear. If there is no such Boolean variable, we can linearize the non-linear pseudo-Boolean inequality directly by expanding the compact pseudo-Boolean inequality set that is equivalent to the non-linear pseudo-Boolean inequality as shown in Proposition 9.4.1.

Given a pseudo-Boolean constraint solver working directly with the compact set representation of pseudo-Boolean constraints, we can stop the linearization process if there is no Boolean variable having at least two occurrences, where at least one of them is non-linear. Instead of linearizing, we add the corresponding compact pseudo-Boolean inequality set of Proposition 9.4.1 directly to the constraint store.

10

PROJECTION

At the end of computation of a CLP($\mathcal{PB}$)-program we have to reformulate the set of collected constraints such that only Boolean variables that appear in the top-level goal remain. The so computed constraint set is the answer constraint of the top-level goal.

Let S be the set of pseudo-Boolean constraints collected at the end of a computation and let V' be the set of Boolean variables occurring in the corresponding top-level goal. We search a set of pseudo-Boolean constraints S' containing only variables in V' such that $\mathsf{Ext}(S')$ equals $\mathsf{Ext}(S)$ restricted to the variables in V' according to the following definition.

Definition 10.0.1 Let S be a set of pseudo-Boolean constraints and let V' be a set of Boolean variables. A *projection* $S|_{V'}$ of S on V' is a set of pseudo-Boolean constraints containing only variables in V' such that

[a] for all $\alpha \in \mathsf{Ext}(S)$ there exists an $\alpha' \in \mathsf{Ext}(S|_{V'})$ with $\alpha(X_i) = \alpha'(X_i)$ for all $X_i \in V'$ and

[b] for all $\alpha' \in \mathsf{Ext}(S|_{V'})$ there exists an $\alpha \in \mathsf{Ext}(S)$ with $\alpha(X_i) = \alpha'(X_i)$ for all $X_i \in V'$.

The first condition of Definition 10.0.1 assures that we do not miss any satisfying solution, whereas the latter assures that all solutions of the reformulated set of pseudo-Boolean constraints are correct. Let V be the set of Boolean variables occurring in S. Then projecting S on V' can also be seen as a *variable elimination* problem. We say we have eliminated the Boolean variables $V \setminus V'$ in the projection $S|_{V'}$ of S.

10.1 PROJECTION FOR PRIME EXTENDED CLAUSES

Projecting a set of prime extended clauses on a set of variables is easy. It suffices to eliminate all variables that need to be projected away for each prime extended clause separately, but that just means to compute strict reductions.

Proposition 10.1.1 *Let S be the set of all prime extended clauses of a set of pseudo-Boolean constraints and V' be a set of Boolean variables. Let*

$$S' := \{L \setminus L' \geq d - |L'| \mid L \geq d \in S \wedge L' = L \cap (V' \cup \overline{V'})\} \tag{10.1}$$

be the set of all strict reductions of the extended clauses in S, where all literals whose variables are not in V' are eliminated. Then S' is a projection of S on V'.

Proof. We need to show Definition 10.0.1 [a] and [b].

[a] Since an extended clause dominates all its reductions, we know that each extended clause in S' is dominated by an extended clause in S and thus S dominates S'. Therefore, $\mathsf{Ext}(S) \subseteq \mathsf{Ext}(S')$ and Definition 10.0.1 [a] is fulfilled.

[b] Suppose that Definition 10.0.1 [b] does not hold. Then there exists an $\alpha' \in \mathsf{Ext}(S')$ such that for all $\alpha \in \mathsf{Ext}(S)$ there is an $X_i \in V'$ with $\alpha(X_i) \neq \alpha'(X_i)$. Therefore, α' is not in the extension of S. Let $L = \{L_i \mid \alpha'(L_i) = 1 \wedge \mathsf{Var}(L_i) \in V'\}$ be the set of literals containing the variables in V' that are mapped to 1 by α'. Since α' is not in the extension of S, we know that fixing the literals in L to 1 in the problem S yields an unsatisfiable set of extended clauses, i.e., $S \cup \{L \geq |L|\}$ is unsatisfiable. Thus, S dominates $L < |L|$ and therefore S dominates $\overline{L} \geq 1$. Since S is a complete set of prime extended clauses, we know that there is an extended clause $L' \geq d' \in S$ that dominates $\overline{L} \geq 1$. Let $L'' \geq d'' \in S'$ be the strict reduction of $L' \geq d'$ obtained by eliminating all literals in $V \cup \overline{V}$. Since none of the literals that are eliminated occur in $\overline{L} \geq 1$, we know that $L'' \geq d''$ dominates $\overline{L} \geq 1$. Since $\alpha'(\overline{L}) = 0$, we have $\alpha' \notin \mathsf{Ext}(\overline{L} \geq 1)$. Thus, $\alpha' \notin \mathsf{Ext}(L'' \geq d'')$ and therefore $\alpha' \notin \mathsf{Ext}(S')$, which is a contradiction. □

With Proposition 10.1.1 the projection on a set of variables for a complete set of prime extended clauses of a set of pseudo-Boolean constraints is computationally no problem. Furthermore, the projection by Proposition 10.1.1 of a set of all prime extended clauses of a set of pseudo-Boolean constraints is the set of all prime extended clauses of the projection of the set of pseudo-Boolean constraints. Hence, we immediately obtain the solved form.

Proposition 10.1.2 *Let S be the set of all prime extended clauses of a set of pseudo-Boolean constraints T and V' be a set of Boolean variables. Let S' be defined by (10.1), then S' is the set of all prime extended clauses of $T|_{V'}$.*

Proof. By Proposition 10.1.1 we know that S' is equivalent to $S|_{V'}$ and thus to $T|_{V'}$. It remains to show that S' consists of all prime extended clauses of $S|_{V'}$. Let $L' \geq d'$ be an extended clause that is dominated by S'. Since S dominates S' and S is the set of all prime extended clauses, we know that there is an extended clause $L \geq d$ in S that dominates $L' \geq d'$. The reduction of $L \geq d$, where we have eliminated all literals in $V' \cup \overline{V'}$, also dominates $L \geq d$, since only literals not occurring in $L' \geq d'$ are eliminated. Since that reduction is in S' by construction, we know that there is a single extended clause in S' that dominates $L' \geq d'$. Since $L' \geq d'$ is arbitrary, we know that any extended clause dominated by S' is dominated by a single extended clause in S'. Thus, S' consists of all prime extended clauses of $S|_{V'}$. □

Given the complete solved form, we see that projection can easily be done. Typically, we do not have the complete solved form, i.e., the set of all prime extended clauses of the set of pseudo-Boolean constraints collected so far. Hence, we need to project arbitrary sets of pseudo-Boolean constraints on a set of Boolean variables. We assume that we have reformulated the set of pseudo-Boolean constraints into an equivalent set of extended clauses. Thus, it suffices to consider the projection of a set of extended clauses onto a set of Boolean variables.

10.2 PROJECTION FOR EXTENDED CLAUSES

Projecting an arbitrary set of extended clauses on a set of Boolean variables is NP-complete[1] and practically a hard problem. We have to build all possible resolvents on the Boolean variables that are to be eliminated. We develop a projection algorithm that is similar to the Fourier-Motzkin variable elimination algorithm for linear constraints over the real numbers.

[1]By eliminating all variables we can decide satisfiability of a set of extended clauses.

Let S be a set of extended clauses and let X be a Boolean variable that needs to be eliminated. The problem is to find a projection $S|_{V\setminus\{X\}}$ of S, where V is the set of all Boolean variables that occur in S. Let

$$\mathbf{res}(S,X) := \left\{ L \geq d \;\middle|\; \begin{array}{l} L \geq d \in \mathsf{SCP}(\mathsf{SRed}(L' + L'' \geq d' + d'')), \\ L' \geq d' \in S, L'' \geq d'' \in S, \\ X \in L' \text{ and } \overline{X} \in L'' \end{array} \right\}, \quad (10.2)$$

then for each possible resolvent on X for any two extended clauses in S, $\mathbf{res}(S,X)$ contains an extended clause that dominates the resolvent by Proposition 7.8.2. Furthermore, all extended clauses in $\mathbf{res}(S,X)$ do neither contain X nor $\overline{X}$.

We show that the set of reductions as defined in Proposition 10.1.1 together with $\mathbf{res}(S,X)$ are a projection of S on $V\setminus\{X\}$. For this purpose we need two lemmas for reasoning on classical clauses.

Lemma 10.2.1 *Let S be a set of extended clauses. We define the set of classical clauses*

$$SC := \{L' \geq 1 \mid L \geq d \in S \wedge L' \subseteq L \wedge |L'| = |L| - d + 1\} .$$

Then $\mathsf{Ext}(S) = \mathsf{Ext}(SC)$ and for each classical clause $L' \geq 1$ that is dominated by S, classical resolution on SC can generate a classical clause that dominates $L' \geq 1$.

Proof. Since SC consists only of reductions of extended clauses in S, we know that S dominates SC. On the other hand, each extended clause of S can be generated by the diagonal sum rule from the classical clauses in SC by construction. Thus, SC dominates S and therefore $\mathsf{Ext}(S) = \mathsf{Ext}(SC)$.

Let $L \geq 1$ be a classical clause that is dominated by S, then SC dominates $L \geq 1$. Since resolution is complete on classical clauses, i.e., can generate all prime implicates of a set of classical clauses [Qui52, Qui55], we know that resolution can generate a classical clause dominating $L \geq 1$. □

Lemma 10.2.2 *Let SC be a set of classical clauses. Let $L \geq 1$ be a classical clause dominated by SC and let X be a Boolean variable such that neither X nor $\overline{X}$ occurs in L. Then $SC' \cup \mathbf{res}(SC,X)$ dominates $L \geq 1$ with $SC' :=$ $\{L' \geq 1 \mid L' \geq 1 \in SC$ and $X, \overline{X} \notin L'\}$.*

Proof. We rely on Tison's method [Tis67] for generating prime implicates of a set of classical clauses[2]. Let V be the set of all variables occurring in the set of classical clauses SC. All prime implicates of SC are generated by the following algorithm due to Tison [Tis67].

$$PI := SC$$
$$\textbf{forall} \quad X \in V \quad \textbf{do} \quad PI := PI \cup \texttt{res}(PI, X) \quad \textbf{endforall}$$

At the end of computation PI is the set of all prime implicates of SC. In the above algorithm PI is always kept in normal form, i.e., PI does not contain classical clauses that are dominated by another classical clause in PI. The correctness proof of Tison's method can be found in [LB82].

Next, we show that $SC' \cup \texttt{res}(SC, X)$ dominates $L \geq 1$. Since SC dominates $L \geq 1$, we know that there is a prime implicate $L' \geq 1$ of SC that dominates $L \geq 1$. Since neither X nor $\overline{X}$ occur in L, we know that this prime implicate $L' \geq 1$ of SC does not contain X or $\overline{X}$. Furthermore, we know that Tison's method will generate $L' \geq 1$. Suppose that we start the loop in Tison's method by resolving on all classical clauses containing X. Note that the order of the variables on which to resolve is arbitrary. After that step we do no longer resolve on X. Therefore all classical clauses that contain X or $\overline{X}$ generate only classical clauses that also contain X or $\overline{X}$. Thus, $L' \geq 1$ must either be already in $SC' \cup \texttt{res}(SC, X)$ or is generated by Tison's method, where all classical clause still containing X or $\overline{X}$ are eliminated. Therefore, we know that $L' \geq 1$ is dominated by $SC' \cup \texttt{res}(SC, X)$. □

Proposition 10.2.3 *Let S be a set of extended clauses. Let X be a Boolean variable and let*

$$S' := \{L \setminus L' \geq d - |L'| \mid L \geq d \in S \text{ and } L' := L \cap \{X, \overline{X}\}\}$$

similar to (10.1). Then $S' \cup \texttt{res}(S, X)$ is a projection of S on $V \setminus \{X\}$.

Proof. We need to show Definition 10.0.1 [a] and [b].

[a] S dominates both S' and $\texttt{res}(S, X)$.

[b] Suppose that Definition 10.0.1 [b] does not hold. Then there exists an $\alpha' \in \mathsf{Ext}(S' \cup \texttt{res}(S, X))$ such that for all assignments $\alpha \in \mathsf{Ext}(S)$ there is an $X_i \in V \setminus X$ with $\alpha(X_i) \neq \alpha'(X_i)$. Therefore, α' is not in the extension of S. Let $L := \{L_i \mid \alpha'(L_i) = 1 \wedge \mathsf{Var}(L_i) \in V \setminus \{X\}\}$ be the set of literals

[2] Tison's method was originally developed for the disjunctive normal form and the consensus operation. By duality all results hold for conjunctive normal form, i.e., classical clauses and resolution.

containing the variables in $V \setminus \{X\}$ that are mapped to 1 by α'. Similar to the proof of Proposition 10.1.1, we derive that S dominates $\overline{L} \geq 1$. By Lemma 10.2.1 we know that there is a classical clause dominating $\overline{L} \geq 1$ that is derivable by classical resolution from the set SC of classical clauses equivalent to S. By Proposition 7.8.2 we know that for each possible resolvent on X for any two extended clauses in S, $\mathtt{res}(S, X)$ contains an extended clause that dominates that resolvent. Let $T = S' \cup \mathtt{res}(S, X)$ and TC be the set of classical clauses that is equivalent to T obtained by applying Lemma 10.2.1 on T. Then by Lemma 10.2.2 we know that the classical clause dominating $\overline{L} \geq 1$ can be derived by classical resolution from TC. Therefore, $S' \cup \mathtt{res}(S, X)$ dominates $\overline{L} \geq 1$. Since $\alpha'(\overline{L}) = 0$, we know that $\alpha' \notin \mathsf{Ext}(S' \cup \mathtt{res}(S, X))$, which is a contradiction. □

With help of Lemma 10.2.1 we transform our problem to a problem for classical clauses, which is then solved using Lemma 10.2.2. Note that the diagonal sum rule need not be applied for calculating a projection.

We obtain a projection of a set of extended clauses S on a set of Boolean variables V' by eliminating one after the other all variables in $V \setminus V'$, where V is the set of Boolean variables occurring in S.

Example 10.2.4 Let us eliminate the Boolean variable C in the set of extended clauses

$$S = \{A+B+C+D \geq 2, A+\overline{C}+\overline{D}+E+F \geq 3, \overline{A}+C+E \geq 1, D+\overline{C}+F \geq 1\} .$$

We build all linear combinations of two extended clauses, where one contains C and the other one contains $\overline{C}$, yielding the set of linear pseudo-Boolean inequalities

$$\{2{\cdot}A+B+E+F \geq 3, 2{\cdot}D+A+B+F \geq 2, 2{\cdot}E+\overline{D}+F \geq 2, \overline{A}+D+E+F \geq 1\} .$$

We transform these linear pseudo-Boolean inequalities into an equivalent set of extended clauses and obtain

$$\mathtt{res}(S, C) = \begin{array}{l} \{A+E \geq 1, A+B \geq 1, A+F \geq 1, A+B+E+F \geq 2, \\ \;B+D+F \geq 1, E+F \geq 1, \overline{D}+E \geq 1\} , \end{array}$$

where dominated extended clauses have been deleted. The set of strict reductions of S, where C and $\overline{C}$ are eliminated, is

$$S' = \{A+B+D \geq 1, A+\overline{D}+E+F \geq 2\} .$$

Since $A+B+D \geq 1$ is already dominated by an extended clause in $\mathbf{res}(S, C)$, we know that

$$S|_{\{A,B,D,E,F\}} = \begin{array}{l}\{A+E \geq 1, A+B \geq 1, A+F \geq 1, \\ A+B+E+F \geq 2, B+D+F \geq 1, \\ E+F \geq 1, \overline{D}+E \geq 1, A+\overline{D}+E+F \geq 2\}\ .\end{array}$$

Next, we eliminate A. Since A occurs only positively in $S|_{\{A,B,D,E,F\}}$, we know that $\mathbf{res}(S|_{\{A,B,D,E,F\}}, A) = \emptyset$. Thus, the projection $S|_{\{B,D,E,F\}}$ consists only of the strict reductions

$$\{E+F \geq 1, B+D+F \geq 1, \overline{D}+E \geq 1\}$$

of $S|_{\{A,B,D,E,F\}}$, where we have eliminated A.

The presented variable elimination method for eliminating a Boolean variable is closely related to the Fourier-Motzkin variable elimination method for linear inequalities over the real numbers [Fou27, Mot36, Sch86]. We briefly recall the Fourier-Motzkin variable elimination method. Let A be a set of linear inequalities over the real numbers. Let V be the set of real variables occurring in A, and let X be a real variable that has to be eliminated. Each inequality can be brought into one of the following three forms

$$-X + \sum_{Y \in V \setminus \{X\}} c_i Y \ \geq \ d \qquad (10.3)$$

$$+X + \sum_{Y \in V \setminus \{X\}} c_i Y \ \geq \ d \qquad (10.4)$$

$$\sum_{Y \in V \setminus \{X\}} c_i Y \ \geq \ d\ . \qquad (10.5)$$

Let $\mathtt{fm}(A, X)$ be the set of all linear inequalities obtained from adding an inequality of the form (10.3) to one of the form (10.4), then $\mathtt{fm}(A, X)$ does no longer contain X. We know that the set of linear inequalities $\mathtt{fm}(A, X)$ plus all inequalities of the form (10.5) describe a projection of A on $V \setminus \{X\}$ [Sch86]. Note that $\mathtt{fm}(A, X)$ corresponds to the resolvents $\mathbf{res}(S, X)$ in our variable elimination method. What is missing is the set of strict reductions in S'. Note that we also have the trivial inequalities $-X_i \geq -1$ and $X_i \geq 0$ for all Boolean variables X_i. Since they are of the form (10.3), resp. (10.4), they need also be considered in the Fourier-Motzkin elimination method. But then we obtain all strict reductions in S'. Thus, our variable elimination method for extended clauses is essentially the Fourier-Motzkin variable elimination method for real constraints, plus the transformation of linear pseudo-Boolean inequalities into an equivalent set of extended clauses. Williams [Wil76] strengthened the Fourier-Motzkin elimination method by certain reduction and rounding operations to work for integer programming problems. Equation (10.2) can be seen as a specialized version of these operations that work for extended clauses. Hooker [Hoo92b] presents a specialized method for classical clauses.

10.3 IMPROVEMENTS

The presented variable elimination method is a computationally expensive task. In the worst case we may generate a quadratic number of linear pseudo-Boolean inequalities that need to be transformed and added to the constraint store for each variable to be eliminated. A possible improvement of the method is to first check whether an extended clause containing a variable that is to be eliminated is dominated by an extended clause not containing that variable.

Let $L + X \geq d \in S$ and suppose that we want to eliminate the Boolean variable X. If S dominates $L \geq d$, we can replace $L + X \geq d$ by $L \geq d$. In the variable elimination process we then need not compute any linear combination with $L + X \geq d$. We can check whether $L + X \geq d$ is redundant with the techniques presented in Section 6.3. We know that S dominates $L \geq d$ if and only if $\texttt{entail\_extendedclause}(S, L \geq d) = (S', \top)$ Similarly, for extended clauses of the form $L + \overline{X} \geq d$. Since a complete test of entailment by **entail_extendedclause** may be computationally too expensive for all extended clauses that contain the variable to be eliminated, we may restrict the test by selecting only some promising extended clauses or limit the computational effort spent for the entailment test. Note that the simplifying resolvents of Section 8.3 have already the property to construct some of these dominated extended clauses. Whether or not a complete entailment test is worthwhile in practice remains to be evaluated.

Typically, we do not only eliminate a single Boolean variable but many Boolean variables. Although we have to eliminate each Boolean variable one after the other, the order is arbitrary and can influence the performance of the variable elimination algorithm. Obviously, it is a good idea to eliminate first all Boolean variables X that occur only positively or only negatively in S, since then $\texttt{res}(S, X) = \emptyset$. As a heuristic rule we propose to choose the Boolean variable that minimizes the number of linear pseudo-Boolean inequalities that are generated by $\texttt{res}(S, X)$. Let $\texttt{occ}(S, L_i) := \{L \geq d \mid L \geq d \in S \text{ and } L_i \in L\}$ be set of all extended clauses in S where the literal L_i occurs. First, we eliminate the Boolean variable X that minimizes

$$|\mathtt{occ}(S, X)| \cdot |\mathtt{occ}(S, \overline{X})| \ .$$

Thus, we minimize the number of generated linear pseudo-Boolean inequalities. We obtain the following Boolean variable elimination algorithm for extended clauses.

Algorithm 10.3.1 *Boolean variable elimination for extended clauses*

```
var_elim(S, V')
   while V' ≠ ∅
      S := ve_simplify(S, V')
      M := min_{Y∈V'}(|occ(S, Y)| · |occ(S, Ȳ)|)
      select X such that |occ(S, X)| · |occ(S, X̄)| = M
      S := {L \ L' ≥ d − |L'| | L ≥ d ∈ S, L' = L ∩ {X, X̄}} ∪ res(S, X)
      V' := V' \ {X}
   endwhile
end var_elim
```

We assume the existence of a function `ve_simplify` that performs some of the simplifications possible by a complete entailment check. Algorithm 10.3.1 calculates a projection S' of S on a set of Boolean variables V' when called as

$$S' := \mathtt{var\_elim}(S, V \setminus V') \ ,$$

where V is the set of Boolean variables occurring in S.

11

CONCLUSION

We have presented a complete collection of techniques for solving pseudo-Boolean constraints in the constraint logic programming language CLP($\mathcal{PB}$). We can use the methods to ensure global consistency (if desired) and to provide an appropriate solved form.

On one hand, pseudo-Boolean constraints are a powerful extension of Boolean constraints. On the other hand, pseudo-Boolean constraints are a restriction of finite domain constraints that allow the fulfillment of the requirements imposed in the context of constraint logic programming, i.e., providing a solved form and ensuring global consistency. The class of treatable problems still covers an important class of typical problems arising in operations research and artificial intelligence.

The set of all prime extended clauses as ideal solved form fits exactly into the context of constraint logic programming. Furthermore, it provides an intuitive and user-friendly understanding because of the natural d out of n interpretation. Since the ideal solved form is typically exponential in size w.r.t. the constraint set, the crucial idea is to approximate the solved form as long as questions imposed by the CLP-environment are not yet answered. The incremental logic cut algorithm of Chapter 6 allows the approximation of the solved form as closely as necessary and furthermore handles entailment questions. Thus, it constitutes the core of our complete pseudo-Boolean constraint solver. The presented simplification techniques enhance the efficiency of the constraint solver and are an attempt to make the theoretical power of cutting plane proofs available for practical use.

A complete pseudo-Boolean constraint solver embedded in CLP($\mathcal{PB}$) encourages a more declarative programming style, since inconsistency can be detected immediately. For many typical problems, i.e., optimization problems, there is no performance penalty when ensuring global consistency. For some CLP-programs efficiency is enhanced or, even better, termination ensured. From a software-engineering point of view it is important that if global consistency is ensured, the program behavior does not change if the underlying constraint solving techniques are modified. In the context of concurrent constraint logic programming the presented constraint solver provides a basic functionality, namely deciding entailment.

11.1 SUMMARY OF RESULTS

All relevant parts of a complete symbolic pseudo-Boolean constraint solver for CLP($\mathcal{PB}$) are covered in this book. The investigation of extended clauses, which are a useful generalization of classical clauses that remains computationally tractable, is a main part of this work. They constitute the solved form of our constraint solver.

The transformation of linear pseudo-Boolean inequalities into an equivalent set of extended clauses is the basic method for handling arbitrary pseudo-Boolean constraints by specialized methods for extended clauses. Encouraging computational results show that the number of extended clauses is typically much smaller than the number of classical clauses needed to represent a linear pseudo-Boolean inequality. Furthermore, the transformation technique is also applicable in the context of classical, polyhedral-based, 0-1 integer programming, i.e., it may serve as a cut generator in a branch-and-cut algorithm.

The deductive system *Generalized Resolution* introduced by Hooker is a complete inference system for linear pseudo-Boolean constraints. We have studied the inference rules of *Generalized Resolution* and specialized them for extended clauses. Applying these inference rules only if they *simplify* the current set of constraints has been shown to carry over the theoretical power of cutting plane proofs into a practical method. We have presented conditions and methods that enable an efficient search for these simplifications. Like the transformation method, the simplification methods can be used as cut generators in the context of classical linear programming based 0-1 integer programming.

The linearization method allows us to handle non-linear pseudo-Boolean constraints in CLP($\mathcal{PB}$). In the context of constraint logic programming a variable elimination algorithm for extended clauses is needed in order to project on the Boolean variables appearing in the top-level goal. We have developed a Fourier-Motzkin like variable elimination algorithm for projecting extended clauses on a set of Boolean variables for this purpose.

The presented pseudo-Boolean constraint solver is a logic-based method for solving pseudo-Boolean problems. As discovered by Hooker [Hoo94b, Hoo94a], logic-based methods and polyhedral methods have many parallels when replacing the polyhedral concepts by their logical counterparts. We have introduced a discrete relaxation, namely pseudo-Boolean unit relaxation, which replaces the linear programming relaxation. Computational results indicate that a pure branch-and-bound algorithm based on pseudo-Boolean unit relaxation is comparably efficient w.r.t. state-of-the-art linear programming based branch-and-bound methods. Furthermore, a pure logic cut algorithm has been given that corresponds to a pure polyhedral cut method for solving linear pseudo-Boolean problems.

One reason for the great success of polyhedral methods in 0-1 integer programming is the number of sophisticated problem-specific polyhedral cut generation methods. Hooker [Hoo94b, Hoo94a] starts the investigation of problem-specific logic cut generation methods. The presented simplification techniques can be seen as general problem independent logic cut generation methods, as needed for a general purpose pseudo-Boolean constraint solver.

11.2 RELATED AND FUTURE RESEARCH

Using logic cut methods in a constraint solver for CLP($\mathcal{PB}$) is the dual approach of the polyhedral cut method for constraint logic programming proposed by Bockmayr [Boc94]. The advantages of polyhedral cutting plane methods in the context of constraint logic programming carry over to logic cut methods. The investigation of strong polyhedral cut generation methods for general linear 0-1 problems and their incorporation into constraint logic programming is currently an active research area.

The investigation of the logic cut approach by Hooker [Hoo94b, Hoo94a] based on the deductive system *Generalized Resolution* [Hoo92a] is a promising alternative to classical linear programming based 0-1 integer programming. But, logic cut and polyhedral cut generation methods do not exclude each other. Transferring methods from one field to the other and comparing both approaches w.r.t. their problem-specific applicability is an interesting topic for future research.

REFERENCES

[AB93] A. Aggoun and N. Beldiceanu. Extending CHIP in order to solve complex scheduling and placement problems. *Mathl. Comput. Modelling*, 17(7):57–73, 1993.

[AK93] H. Aït-Kaci. An introduction to LIFE – programming with logic, inheritance, functions and equations. In D. Miller, editor, *Proceedings of the International Symposium on Logic Programming*, pages 52–68, Vancouver, Canada, 1993. MIT Press.

[ASS+88] A. Aiba, K. Sakai, Y. Sato, D. J. Hawley, and R. Hasegawa. Constraint logic programming language CAL. In *Proceedings International Conference on Fifth Generation Computer Systems*, pages 263–276, Tokyo, Japan, 1988. Springer-Verlag.

[Bar92a] P. Barth. CLP($\mathcal{PB}$) A Meta-interpreter in CLP($\mathcal{R}$). Technical Report MPI-I-92-233, Max-Planck-Institut für Informatik, Saarbrücken, 1992.

[Bar92b] P. Barth. A complete symbolic 0-1 constraint solver. In *ACCLAIM Kick-Off Workshop*, Stockholm, Sweden, November 1992.

[Bar93a] P. Barth. A complete symbolic 0-1 constraint solver. In A. Bockmayr and F. J. Radermacher, editors, *Künstliche Intelligenz und Operations Research, Workshop, Extended Abstracts*, MPI-I-93-234, pages 11–13, Berlin, Germany, September 1993.

[Bar93b] P. Barth. A complete symbolic 0-1 constraint solver. In F. Benhamou, A. Colmerauer, and G. Smolka, editors, *Third Workshop on Constraint Logic Programming WCLP*, Marseille, France, March 1993.

[Bar93c] P. Barth. Linear 0-1 inequalities and extended clauses. In *Proceedings International Conference on Logic Programming and Automated Reasoning LPAR*, pages 40–51, St. Petersburg, Russia, July 1993. Springer, LNAI 698.

[Bar94] P. Barth. Linear 0-1 inequalities and extended clauses. Technical Report MPI-I-94-216, Max-Planck-Institut für Informatik, Saarbrücken, April 1994.

[Bar95] P. Barth. A Davis-Putnam based enumeration algorithm for linear pseudo-Boolean optimization. Technical Report MPI-I-95-2-003, Max-Planck-Institut für Informatik, Saarbrücken, January 1995. System available:
`ftp.mpi-sb.mpg.de:pub/guide/staff/barth/opbdp/opbdp.tar.Z`,
`http://www.mpi-sb.mpg.de/~barth/opbdp/opbdp.html` .

[BB93] P. Barth and A. Bockmayr. Solving 0-1 problems in CLP($\mathcal{PB}$). In *Proceedings 9th Conference on Artificial Intelligence for Applications CAIA*, pages 263–269, Orlando, Florida, 1993. IEEE.

[BB95] P. Barth and A. Bockmayr. Finite domain and cutting plane techniques in CLP($\mathcal{PB}$). In L. Sterling, editor, *Proceedings 12th International Conference on Logic Programming ICLP*, pages 133–147, Kanagawa, Japan, 1995. MIT Press.

[BBI92] R. E. Bixby, E. A. Boyd, and R. Indovina. MIPLIB: A Test Set of Mixed-Integer Programming Problems. *SIAM News 25*, page 16, 1992. ftp: softlib.cs.rice.edu.

[BC94] N. Beldiceanu and E. Contejean. Introducing global constraints in CHIP. *Mathl. Comput. Modelling*, 20(12):97–123, 1994.

[BCC93a] E. Balas, S. Ceria, and G. Cornuéjols. A lift-and-project cutting plane algorithm for mixed 0-1 programs. *Mathematical Programming*, 58:295–324, 1993.

[BCC93b] E. Balas, S. Ceria, and G. Cornuéjols. Solving mixed 0-1 programs by a lift-and-project method. In *Symposium on Discrete Algorithms SODA*, Austin, Texas, 1993. ACM - SIAM.

[BCC94] E. Balas, S. Ceria, and G. Cornuéjols. Mixed 0–1 programming by lift-and-project in a branch-and-cut framework. Technical report, GSIA, Carnegie Mellon University, February 1994.

[BDP93] P. Boizumault, Y. Delon, and L. Péridy. Solving a real-life planning exams problem using constraint logic programming. In M. Meyer, editor, *Constraint Processing: Proceedings of the International Workshop at CSAM'93*, Research Report RR-93-39, pages 107–112, St. Petersburg, Russia, July 1993. DFKI Kaiserslautern.

[Bea85] J. Beasly. Bounds for two-dimensional cutting. *Operational Research Society*, 36(1):71–74, 1985.

[Ben93] F. Benhamou. Boolean algorithms in Prolog III. In F. Benhamou and A. Colmerauer, editors, *Constraint Logic Programming – Selected Research*, chapter 17, pages 307–326. MIT Press, 1993.

[Ber88] F. Berthier. Managing underlying assumptions of a financial planning model in CHIP. Technical Report TR-LP-39, ECRC, Munich, Germany, 1988.

[Ber90] F. Berthier. Solving financial decision problems with CHIP. In J.-L. Le Moigne and P. Bourgine, editors, *CECOIA 2: Proceedings 2nd Conference on Economics and Artificial Intelligence*, pages 233–238, Paris, France, 1990.

[BJL86] C. E. Blair, R. G. Jeroslow, and J. K. Lowe. Some results and experiments in programming techniques for propositional logic. *Computers and Operations Research*, 13(5):633–645, 1986.

[BL95] S. Breitinger and H. C. R. Lock. Using constraint logic programming for industrial scheduling problems. In C. Beierle and L. Plümer, editors, *Logic Programming: Formal Methods and Practical Applications*, Studies in Computer Science and Artificial Intelligence, chapter 9, pages 273–299. Elsevier, 1995.

[BLPN95] P. Baptiste, C. Le Pape, and W. Nuijten. Incorporating efficient operations research algorithms in constraint-based scheduling. In J. N. Hooker and M. Ginsberg, editors, *Proceedings First International Workshop on Artificial Intelligence and Operations Research*, Oregon, 1995.

[BM84a] E. Balas and J. B. Mazzola. Nonlinear 0-1 programming: I. Linearization techniques. *Mathematical Programming*, 30:1–21, 1984.

[BM84b] E. Balas and J. B. Mazzola. Nonlinear 0-1 programming: II. Dominance relations and algorithms. *Mathematical Programming*, 30:22–45, 1984.

[BMvH94] F. Benhamou, D. McAllester, and P. van Hentenryck. CLP(Intervals) revisited. In M. Bruynooghe, editor, *Proceedings International Symposium on Logic Programming ILPS*, pages 124–138, Santa Margherita Ligure, Italy, 1994. MIT Press.

[BO93] F. Benhamou and W. Older. Applying interval arithmetic to real, integer and Boolean constraints. *Logic Programming: The ALP Newsletter*, 6(2):13–14, 1993. (Extended Abstract).

[Boc91] A. Bockmayr. Logic programming with pseudo-Boolean constraints. Technical Report MPI-I-91-227, Max-Planck-Institut für Informatik, Saarbrücken, 1991.

[Boc92a] A. Bockmayr. 0-1 constraints and 0-1 optimization. In *ACCLAIM Kick-Off Workshop*, Stockholm, Sweden, November 1992.

[Boc92b] A. Bockmayr. Embedding OR techniques in constraint logic programming. In *Operations Research '92. 17th Symposium on Operations Research*, Hamburg, Germany, 1992.

[Boc93a] A. Bockmayr. 0-1 constraints and 0-1 optimization. In F. Benhamou, A. Colmerauer, and G. Smolka, editors, *Third Workshop on Constraint Logic Programming WCLP*, Marseille, France, March 1993.

[Boc93b] A. Bockmayr. Logic programming with pseudo-Boolean constraints. In F. Benhamou and A. Colmerauer, editors, *Constraint Logic Programming - Selected Research*, chapter 18, pages 327–350. MIT Press, 1993.

[Boc93c] A. Bockmayr. Using strong cutting planes in constraint logic programming. In *Operations Research '93. 18th Symposium on Operations Research*, Köln, Germany, 1993.

[Boc94] A. Bockmayr. Cutting planes in constraint logic programming. Technical Report MPI-I-94-207, Max-Planck-Institut für Informatik, Saarbrücken, Saarbrücken, February 1994.

[Boc95] A. Bockmayr. Solving pseudo-Boolean constraints. In A. Podelski, editor, *Constraint Programming: Basics and Trends*, pages 22–38. Springer, LNCS 910, 1995.

[BR89] A. Bockmayr and H. H. Rath. Algorithms for Boolean unification. In *Proceedings FAW Workshop on Boolean Functions, Propositional Logic, and AI Systems*. FAW Ulm, Germany, 1989.

[Bry86] R. E. Bryant. Graph-based algorithms for Boolean function manipulation. *IEEE Transactions on Computers*, C-35:677–691, 1986.

[Bry92] R. E. Bryant. Symbolic Boolean manipulation with ordered binary-decision diagrams. *ACM Computing Surveys*, 24(3):293–318, 1992.

[Brz91] C. Brzoska. Temporal logic programming and its relation to constraint logic programming. In V. Saraswat and K. Ueda, editors, *Proceedings International Logic Programming Symposium ILPS*, pages 661–677, San Diego, 1991. MIT Press.

[BS87] W. Büttner and H. Simonis. Embedding Boolean expressions in logic programming. *Journal of Symbolic Computation*, 4(2):191–205, 1987.

[BS92] B. Benhamou and L. Sais. Theoretical study of symmetries in propositional calculus and applications. In D. Kapur, editor, *Proceedings 11th International Conference on Automated Deduction CADE*, pages 281–294, Saratoga Springs, NY, 1992. Springer, LNCS 607.

[CCT87] W. Cook, C. R. Coullard, and G. Turán. On the complexity of cutting plane proofs. *Discrete Applied Mathematics*, 18:25–38, 1987.

[CD93] P. Codognet and D. Diaz. Boolean constraint solving using clp(FD). In D. Miller, editor, *Proceedings International Symposium on Logic Programming*, pages 525–529, 1993.

[CD94] P. Codognet and D. Diaz. clp(B): Combining simplicity and efficiency in Boolean constraint solving. In M. Hermenegildo and J. Penjam, editors, *Programming Language Implementation and Logic Programming PLILP*, pages 244–260, Madrid, Spain, 1994. Springer, LNCS 844.

[CFGG95] A. Chamard, A. Fischler, B. Guinaudeau, and A. Guillaud. CHIC lessons on CLP methodology. Technical Report ESPRIT EP5291 / D2.1.2.3, CHIC Project, Dassault Aviation, Bull, 1995.

[CG93] M. Carlsson and M. Grindal. Automatic frequency assignment for cellular telephones using constraint satisfaction techniques. In D. S. Warren, editor, *Proceedings 10th International Conference on Logic Programming ICLP*, pages 647–665, Budapest, Hungary, 1993. MIT Press.

[Chv73] V. Chvátal. Edmund polytopes and a hierarchy of combinatorial problems. *Discrete Mathematics*, 4:305–337, 1973.

[CJP83] H. Crowder, E. L. Johnson, and M. Padberg. Solving large-scale zero-one linear programming problems. *Operations Research*, 31(5):803–834, September 1983.

[CKPR73] A. Colmerauer, H. Kanouia, R. Pasero, and P. Roussel. Un système de communication homme-machine en français. Technical report, Université d'Aix-Marseille II, Groupe Intelligence Artificielle, 1973.

[CL94] C. K. Chiu and J. H. M. Lee. Towards practical interval constraint solving in logic programming. In M. Bruynooghe, editor, *Proceedings International Symposium on Logic Programming ILPS*, pages 109–123, Santa Margherita Ligure, Italy, 1994. MIT Press.

[CM87] W. F. Clocksin and C. S. Mellish. *Programming in Prolog.* Springer, 1987. Third revised and extended edition.

[Col75] G. E. Collins. Quantifier elimination for the elementary theory of real closed fields by cylindrical algebraic decomposition. In H. Brakhage, editor, *Automata Theory and Formal Languages, 2nd GI Conference*, volume 33, pages 134–183, Kaiserslautern, West Germany, 1975. Springer, LNCS 33.

[Col86] A. Colmerauer. Note sur Prolog III. In *Actes du Seminaire 1986 - Programmation en Logique*, pages 159–173, Tregastel, France, Mai 1986. CNET.

[Col87] A. Colmerauer. Introduction to PROLOG III. In *Fourth Annual ESPRIT Conference*, Bruxelles, 1987. North Holland.

[Col90] A. Colmerauer. An introduction to Prolog III. *Communications of the ACM*, 33(7):69–90, 1990.

[Coo71] S. A. Cook. The complexity of theorem proving procedures. In *Proceedings Third ACM Symposium on the Theory of Computing*, pages 151–158. ACM, 1971.

[Cra93] J.-I. Cras. A review of industrial constraint satisfaction tools. AI perspectives report, AI Intelligence, Oxford, 1993.

[CW77] N. Christofides and C. Whitlock. An algorithm for two-dimensional cutting problems. *Operations Research*, 25(1):30–44, 1977.

[CW90] M. Carlsson and J. Widen. *SICStus Prolog Users Manual.* Swedish Institute of Computer Science, 1990. SICS/R-88/88007C.

[Dan57] G. B. Dantzig. Discrete variable extremum problems. *Operations Research*, 5:266–277, 1957.

[Dan60] G. B. Dantzig. On the significance of solving linear programming problems with some integer variables. The Rand Corporation document P1486 (1958), 1960. also published in *Econometrica* 28 (1961), 30–44.

[DC93] D. Diaz and P. Codognet. A minimal extension of the wam for clp(fd). In D. S. Warren, editor, *Proceedings 10th International Conference on Logic Programming ICLP*, pages 774–792, Budapest, Hungary, 1993. MIT Press.

[DEC93] B. L. Dietrich, L. F. Escudero, and F. Chance. Efficient reformulation for 0-1 programs – Methods and computational results. *Discrete Applied Mathematics*, 42:147–175, 1993.

[DP60] M. Davis and H. Putnam. A computing procedure for quantification theory. *Journal of the ACM*, 7:201–205, 1960.

[DSvH88a] M. Dincbas, H. Simonis, and P. van Hentenryck. Solving a cutting-stock problem in constraint logic programming. In R. Kowalski and K. Bowen, editors, *Proceedings Fifth International Conference and Symposium on Logic Programming*, pages 42–58, Seattle, Washington, 1988. MIT Press.

[DSvH88b] M. Dincbas, H. Simonis, and P. van Hentenryck. Solving large scheduling problems in logic programming. In *Proceedings EURO-TIMS Joint International Conference on Operations Research and Management Science*, Paris, France, 1988.

[DSvH88c] M. Dincbas, H. Simonis, and P. van Hentenryck. Solving the car sequencing problem in constraint logic programming. In *Proceedings European Conference on Artificial Intelligence ECAI*, pages 290–295, Munich, Germany, 1988.

[DSvH90] M. Dincbas, H. Simonis, and P. van Hentenryck. Solving large combinatorial problems in logic programming. *Journal of Logic Programming*, 8(1–2):75–93, 1990.

[DvHS+88a] M. Dincbas, P. van Hentenryck, H. Simonis, A. Aggoun, and T. Graf. The constraint logic programming language CHIP. In *Proceedings International Conference on Fifth Generation Computer Systems*, pages 693–702, Tokyo, Japan, 1988. Springer-Verlag.

[DvHS+88b] M. Dincbas, P. van Hentenryck, H. Simonis, A. Aggoun, and T. Graf. Applications of CHIP to industrial and engineering problems. In *Proceedings First International Conference on Industrial and Engineering Applications of Artificial Intelligence and Expert Systems*, Tullahoma, Tennessee, 1988.

[Fag93] F. Fages. On the semantics of optimization predicates in CLP languages. In P. Kanellakis, J.-L. Lassez, and V. Saraswat, editors, *First International Workshop on Principles and Practice of Constraint Programming PPCP*, Newport, Rhode Island, 1993.

[FGGB66] R. J. Freeman, D. C. Gogerty, G. W. Graves, and R. B. S. Brooks. A mathematical model of supply support for space operations. *Operations Research*, 14:1–15, 1966.

[FGK90] R. Fourer, D. M. Gay, and B. W. Kernighan. A modelling language for mathematical programming. *Management Science*, 36:519–554, 1990.

[FHJ91] T. Franzén, S. Haridi, and S. Janson. An overview of the andorra kernel language. In L.-H. Eriksson, L. Hallnäs, and P. Schroeder-Heister, editors, *Proceedings Second International Workshop on Extensions of Logic Programming ELP*, pages 163–180, Stockholm, Sweden, 1991. Springer, LNAI 596.

[For59] R. Fortet. L'algèbre de Boole et ses applications en recherche opérationelle. *Cahiers du Centre d'Études de Recherche Opérationelle*, 1(4):5–36, 1959.

[For60] R. Fortet. Applications de l'algèbre de Boole en recherche opérationelle. *Revue Française d'Informatique et de Recherche Opérationelle*, 4(14):17–25, 1960.

[Fou27] J. B. Fourier. [reported in:] Analyse de travaux de l'Académie Royale des Sciences, pendant l'année 1824, Partie mathématique; Histoire de l'Académie Royale des Sciences de l'Institut de France, 1827. Partial English Translation in *Translation of a Report by Fourier on his work on linear inequalities*, Operations Research 10 (1973) 38–42.

[Ger94] C. Gervet. Conjuncto: Constraint logic programming with finite set domains. Technical Report ECRC-94-15, ECRC, 1994.

[GG80] D. Granot and F. Granot. Generalized covering relaxation for 0-1 programs. *Operations Research*, 28:1442–1450, 1980.

[GH71] F. Granot and P. L. Hammer. On the use of Boolean functions in 0-1 programming. *Methods of Operations Research*, 12:154–184, 1971.

[GH75] F. Granot and P. L. Hammer. On the role of generalized covering problems. *Cahiers du Centre d'Études de Recherche Opérationelle*, 17:277–289, 1975.

[GJ79] M. R. Garey and D. S. Johnson. *Computers and Intractability: A Guide to the Theory of NP-Completeness.* Freeman, San Francisco, CA, 1979.

[Gom58] R. E. Gomory. Outline of an algorithm for integer solutions to linear programs. *Bull. AMS*, 64:275–278, 1958.

[GU89] G. Gallo and G. Urbani. Algorithms for testing the satisfiability of propositional formulae. *Journal of Logic Programming*, 7:45–61, 1989.

[GW73] F. Glover and E. Wolsey. Further reduction of zero-one polynomial programs to zero-one linear programming problems. *Operations Research*, 21(1):156–161, 1973.

[Hak85] A. Haken. The intractability of resolution. *Theoretical Computer Science*, 39(2 & 3):297–308, 1985.

[Hal35] P. Hall. On representatives of subsets. *Journal of the London Mathematical Society*, 10:26–30, 1935.

[Ham64] P. L. Hammer. Systems of pseudo-Boolean equations and inequalities. *Bulletin de l'Academie Polonaise des Sciences-Serie des Sciences Math. , Astr. et Phys.*, XII(11):673–680, 1964.

[Ham68] P. L. Hammer. Plant location – a pseudo-Boolean approach. *Israel Journal of Technology*, pages 350–362, 1968.

[HF90] J. N. Hooker and C. Fedjki. Branch-and-cut solution of inference problems in propositional logic. *Annals of Mathematics and Artificial Intelligence*, 1:123–139, 1990.

[HHT94] F. Harche, J. N. Hooker, and G. L. Thompson. A computational study of satisfiability algorithms for propositional logic. *ORSA Journal on Computing*, 6(4):423–435, 1994.

[HJ90] S. Haridi and S. Janson. Kernel Andorra Prolog and its computation model. In D. H. Warren and P. Szeredi, editors, *Proceedings 7th International Conference on Logic Programming ICLP*, pages 31–48, Jerusalem, Israel, 1990. MIT Press.

[HJM+92] N. C. Heintze, J. Jaffar, S. Michaylov, P. J. Stuckey, and R. Yap. The CLP($\mathcal{R}$) programmer's manual - version 1.2. Technical report, IBM Thomas J. Watson Research Center, September 1992.

[HJM93] P. Hansen, B. Jaumard, and V. Mathon. Constrained nonlinear 0-1 programming. *ORSA Journal on Computing*, 5(2):97–119, 1993.

[HJO+92] W. Havens, J. Jones, R. Ovans, S. Sidebottom, and G. Sidebottom. Echidna: A constraint logic programming shell. In *Proceedings Pacific Rim Conference on AI*, pages 165–171, 1992.

[HL88] T. Huynh and C. Lassez. A CLP($\mathcal{R}$) options trading analysis system. In R. Kowalski and K. Bowen, editors, *Proceedings Fifth International Conference and Symposium on Logic Programming*, pages 59–69, Seattle, Washington, 1988. MIT Press.

[HMS87] N. Heintze, S. Michaylov, and P. Stuckey. CLP($\mathcal{R}$) and some electrical engineering problems. In J.-L. Lassez, editor, *Proceedings 4th International Conference on Logic Programming ICLP*, pages 675–703, Melbourne, Australia, May 1987. MIT Press.

[HMSY89] N. C. Heintze, S. Michaylov, P. J. Stuckey, and R. Yap. On meta-programming in CLP($\mathcal{R}$). In E. Lusk and R. Overbeek, editors, *Proceedings of the North American Conference on Logic Programming*, pages 52–68. MIT Press, October 1989.

[Hol95] C. Holzbaur. OFAI clp(q,r) manual, edition 1.3.3. Technical Report TR-95-09, Austrian Research Institute for Artificial Intelligence, Vienna, 1995.

[Hon90] H. Hong. *Improvements in CAD-Based Quantifier Elimination.* PhD thesis, Ohio State University, 1990.

[Hon92] H. Hong. Non-linear constraints solving over real numbers in constraint logic programming (introducing RISC-CLP). RISC-Linz Report Series 92-08, Research Institute for Symbolic Computation,Johannes Kepler Institute, A-4040 Linz, Austria, Europe, January 1992.

[Hoo88a] J. N. Hooker. A quantitative approach to logical inference. *Decision Support Systems*, 4:45–69, 1988.

[Hoo88b] J. N. Hooker. Resolution vs. cutting plane solution of inference problems: some computational experience. *Operations Research Letters*, 7(1):1–7, 1988.

[Hoo89] J. N. Hooker. Input proofs and rank one cutting planes. *ORSA Journal on Computing*, 1:137–145, 1989.

[Hoo92a] J. N. Hooker. Generalized resolution for 0-1 linear inequalities. *Annals of Mathematics and Artificial Intelligence*, 6:271–286, 1992.

[Hoo92b] J. N. Hooker. Logical inference and polyhedral projection. *ORSA Journal on Computing*, 1(3):137–145, 1992.

[Hoo93] J. N. Hooker. Solving the incremental satisfiability problem. *Journal of Logic Programming*, 15:177–186, 1993.

[Hoo94a] J. N. Hooker. Logic-based methods for optimization. *ORSA CSTS Newsletter*, 15(2):4–11, 1994.

[Hoo94b] J. N. Hooker. Logic-based methods for optimization: A tutorial. Working Paper 1994-05, GSIA, Carnegie-Mellon-University, 1994.

[HR68] P. L. Hammer and S. Rudeanu. *Boolean Methods in Operations Research and Related Areas*. Springer-Verlag, 1968.

[HRR63] P. L. Hammer, I. Rosenberg, and S. Rudeanu. Application of discrete linear programming to the minimization of Boolean functions. *Rev. Math. Pures Appl.*, 8:459–475, 1963. (in Russian).

[HV94] J. N. Hooker and V. Vinay. Branching rules for satisfiability. In *Third International Symposium on Artificial Intelligence and Mathematics*, Fort Lauderdale, Florida, January 1994. to appear in Journal of Automated Reasoning.

[IFP95] IF-prolog 5.0, 1995. IF Computer GmbH, Siemens/Nixdorf.

[Jac92] P. Jackson. Computing prime implicates incrementally. In D. Kapur, editor, *Proceedings 11th International Conference on Automated Deduction CADE*, pages 253–267, Saratoga Springs, NY, 1992. Springer, LNCS 607.

[JKS85] E. L. Johnson, M. M. Kostreva, and U. H. Suhl. Solving 0-1 integer programming problems arising from large scale planning models. *Operations Research*, 33(4):803–819, July 1985.

[JL87] J. Jaffar and J.-L. Lassez. Constraint logic programming. In *Proceedings 14th ACM Symposium on Principles of Programming Languages*, pages 111–119, München, Germany, January 1987. ACM.

[JM94] J. Jaffar and M. J. Maher. Constraint logic programming: A survey. *Journal of Logic Programming*, 19/20:503–581, 1994.

[JMSY90] J. Jaffar, S. Michaylov, P. Stuckey, and R. Yap. The CLP($\mathcal{R}$) language and system. Technical Report RC 16292 (#72336) 11/15/90, IBM Research Division, November 1990.

[JP90] P. Jackson and J. Pais. Computing prime implicants. In M. Stickel, editor, *Proceedings 10th International Conference on Automated Deduction CADE, Kaiserslautern*, pages 543–557. Springer, LNCS 449, 1990.

[JW90] R. G. Jeroslow and J. Wang. Solving propositional satisfiability problems. *Annals of Mathematics and AI*, 1:167–187, 1990.

[Kar72] R. M. Karp. Reducibility among combinatorial problems. In R. E. Miller and J. W. Thatcher, editors, *Complexity of Computer Computations*, pages 85–103. Plenum Press, New York, 1972.

[Kar88] K. Karplus. Representing Boolean functions with if-then-else DAGs. Technical Report UCSC-CRL-88-28, Computer Research Laboratory, Univ. of California, Santa Cruz, Nov. 1988.

[KGV83] S. Kirkpatrick, C. D. Gelatt, and M. P. Vecchi. Optimization by simulated annealing. *Science*, 220:671–680, 1983.

[Kow74] R. A. Kowalski. Predicate logic as programming language. In *Proceedings of the IFIP Congress*, pages 569–574, Amsterdam, Netherlands, 1974. North Holland.

[KT90] A. Kean and G. Tsiknis. An incremental method for generating prime implicants/implicates. *Journal of Symbolic Computation*, 9:185–206, 1990.

[Kum92] V. Kumar. Algorithms for constraint-satisfaction problems: A survey. *A.I. Magazine*, 13(1):32–44, 1992.

[LB82] M. C. Loui and G. Bilardi. The correctness of Tison's method for generating prime implicants/implicates. Technical Report R-952, UILU-ENG 82-2218, Coordinated Science Laboratory, University of Illinois at Urbana-Champaign, 1982.

[LMY87] C. Lassez, K. McAloon, and R. Yap. Constraint logic programming and options trading. *IEEE Expert, Special Issue on Financial Software*, 2(3):42–50, 1987.

[Lov78] D. W. Loveland. *Automated theorem proving : A logical basis*, volume 6 of *Fundamental studies in computer science*. North-Holland, Amsterdam, 1978.

[LP94] C. Le Pape. Implementation of resource constraints in Ilog schedule: A library for the development of constraint-based scheduling systems. *Intelligent Systems Engineering*, 3:55–66, 1994.

[Mac77] A. K. Mackworth. Consistency in networks of relations. *Artificial Intelligence*, 8:99–118, 1977.

[MAC+89] M. Meier, A. Aggoun, D. Chan, P. Dufresne, R. Enders, D. H. de Villeneuve, A. Herold, P. Kay, B. Perez, E. van Rossum, and J. Schimpf. SEPIA – an extendible Prolog system. In *Proceedings 11th Word Computer Congress IFIP'89*, pages 1127–1132, San Francisco, 1989.

[MB93] A. Meeraus and A. Brooke. *GAMS: A User's Guide*. Boyd and Fraser publishing, 1993.

[MN86] U. Martin and T. Nipkow. Unification in Boolean rings. In J. H. Siekmann, editor, *Proceedings 8th International Conference on Automated Deduction CADE*, pages 506–513, Oxford, UK, 1986. Springer, LNCS 230.

[MN88] U. Martin and T. Nipkow. Unification in Boolean rings. *Journal of Automated Reasoning*, 4:381–396, 1988.

[MN89] U. Martin and T. Nipkow. Boolean unification – the story so far. *Journal of Symbolic Computation*, 7:275–293, 1989.

[Mot36] T. S. Motzkin. *Beiträge zur Theorie der linearen Ungleichungen*. PhD thesis, (Inaugural Dissertation Basel), Azriel, Jerusalem, 1936.

[MR91] I. Mitterreiter and F. J. Radermacher. Experiments on the running time behavior of some algorithms solving propositional logic problems. Technical report, FAW Ulm, 1991.

[MS85] B. Monien and E. Speckenmeyer. Solving satisfiability in less than 2^n steps. *Discrete Applied Mathematics*, 10:287–295, 1985.

[MS94] K. Marriott and P. J. Stuckey. Semantics of constraint logic programs with optimization. In *Integration of Declarative Paradigms. Proceedings ICLP'94 Postconference Workshop*, Santa Margherita Ligure, Italy, June 1994.

[MW89] K. I. M. McKinnon and H. P. Williams. Constructing integer programming models by the predicate calculus. *Annals of Operations Research*, 21:227–246, 1989.

[NSS94] G. L. Nemhauser, M. W. P. Savelsbergh, and G. Sigismondi. MINTO, a Mixed INTeger Optimizer. *Operations Research Letters*, 15:47–58, 1994.

[NW88] G. L. Nemhauser and L. A. Wolsey. *Integer and Combinatorial Optimization.* Series in Discrete Mathematics and Optimization. Wiley–Interscience, 1988.

[OB93] W. Older and F. Benhamou. Programming in CLP(BNR). In P. Kanellakis, J.-L. Lassez, and V. Saraswat, editors, *First International Workshop on Principles and Practice of Constraint Programming PPCP*, Newport, Rhode Island, 1993.

[OMT89] A. Oplobedu, J. Marcovitch, and Y. Tourbier. Charme: Un langage industriel de programmation par contraintes, illustré par une application chez Renault. In *9th International Workshop on Expert Systems and their Applications: General Conference*, volume 1, pages 55–70, Avignon, France, May 29th – June 2nd 1989.

[OV90] W. Older and A. Vellino. Extending Prolog with constraint arithmetic on real intervals. In *Proceedings Canadian Conference on Electrical and Computer Engineering*, pages 14.1.1–14.1.4, 1990.

[PR91] M. Padberg and G. Rinaldi. A branch-and-cut algorithm for the resolution of large-scale symmetric traveling salesman problems. *SIAM Review*, 33(1):60–100, 1991.

[PS95] A. Podelski and G. Smolka. Operational semantics of constraint logic programming with coroutining. In L. Sterling, editor, *Proceedings 12th International Conference on Logic Programming ICLP*, pages 449–463, Kanagawa, Japan, 1995. MIT Press.

[Pug94] J.-F. Puget. A C++ implementation of CLP. In *Proceedings Second Singapore International Conference on Intelligent Systems*, Singapore, 1994.

[PWW69] A. B. Pritsker, L. J. Watters, and P. M. Wolfe. Multiproject scheduling with limited resources: a zero-one integer programming approach. *Management Science*, 16(1):93–108, 1969.

[Qui52] W. V. Quine. The problem of simplifying truth functions. *American Mathematical Monthly*, 59:521–531, 1952.

[Qui55] W. V. Quine. A way to simplify truth functions. *American Mathematical Monthly*, 62:627–631, 1955.

[Rhy70] J. Rhys. A selection problem of shared fixed costs and networks. *Management Science*, 17:200–207, 1970.

[Rob65] J. A. Robinson. A machine-oriented logic based on the resolution principle. *Journal of the ACM*, 12(1):23–41, 1965.

[Sar88] V. A. Saraswat. A somewhat logical formulation of CLP synchronization primitives. In R. Kowalski and K. Bowen, editors, *Proceedings Fifth International Conference and Symposium on Logic Programming*, pages 1298–1314, Seattle, Washington, 1988. MIT Press.

[Sar89] V. A. Saraswat. *Concurrent Constraint Programming Languages.* PhD thesis, Carnegie-Mellon University, 1989. Revised version appears as [Sar93].

[Sar93] V. A. Saraswat. *Concurrent Constraint Programming.* MIT Press, 1993.

[Sch86] A. Schrijver. *Theory of Linear and Integer Programming.* John Wiley, 1986.

[SD90] H. Simonis and M. Dincbas. Propositional calculus problems in CHIP. Technical report TR-LP-48, ECRC, May 1990. full version of [SD93].

[SD93] H. Simonis and M. Dincbas. Propositional calculus problems in CHIP. In F. Benhamou and A. Colmerauer, editors, *Constraint Logic Programming - Selected Research*, chapter 15, pages 269–285. MIT Press, 1993.

[SHW93a] G. Smolka, M. Henz, and J. Würtz. Object-oriented concurrent constraint programming in Oz. Research Report RR-93-16, Deutsches Forschungszentrum für Künstliche Intelligenz, April 1993. Previous version appeared as [SHW93b].

[SHW93b] G. Smolka, M. Henz, and J. Würtz. Oz – a programming language for multi-agent systems. In *Proceedings International Joint Conference on Artificial Intelligence*, pages 404–409, Chambéry, France, 1993. Morgan Kaufmann Publishers.

[Smo94] G. Smolka. A calculus for higher-order concurrent constraint programming with deep guards. Research Report RR-94-03, Deutsches Forschungszentrum für Künstliche Intelligenz, February 1994.

[SND88] H. Simonis, H. Nguyen, and M. Dincbas. Verification of digital circuits using CHIP. In G. J. Milne, editor, *Proceedings of the IFIP WG 10.2 International Working conference on the Fusion of Hardware Design and Verification*, Glasgow, Scotland, July 1988. IFIP, North-Holland.

[SSW94] C. Schulte, G. Smolka, and J. Würtz. Encapsulated search and constraint programming in Oz. In A. Borning, editor, *Second International Workshop on Principles and Practice of Constraint Programming PPCP*, pages 134–150, Orcas Island, Seattle, 1994. Springer, LNCS 874.

[Tah71] H. A. Taha. Sequencing by implicit ranking and zero-one polynomial programming. *AIIE Transactions*, 3:299–308, 1971.

[Tar51] A. Tarski. *A Decision Method for Elementary Algebra and Geometry.* University of California Press, Berkeley and Los Angeles, 1951.

[Tis67] P. Tison. Generalized consensus theory and application to the minimization of Boolean functions. *IEEE Transaction on Electronic Computers*, EC-16(4):446–456, 1967.

[Tri93] M. Trick. The second dimacs international algorithm implementation challenge: Clique, graph coloring, and satisfiability. ftp: dimacs.rutgers.edu, 1993.

[Tsa93] E. Tsang. *Foundations of Constraint Satisfaction.* Academic Press, 1993.

[vG84] A. van Gelder. A satisfiability tester for non-clausal propositional calculus. In R. E. Shostak, editor, *Proceedings 7th International Conference on Automated Deduction CADE*, pages 101–112, Napa, CA, May 1984. Springer, LNCS 170.

[vH89a] P. van Hentenryck. *Constraint satisfaction in logic programming.* MIT Press, 1989.

[vH89b] P. van Hentenryck. A logic language for combinatorial optimization. *Annals of Operations Research*, 21:247–274, 1989.

[vH92] P. van Hentenryck. Scheduling and packing in the constraint language cc(FD). Technical Report CS-92-43, Department of Computer Science, Brown University, September 1992.

[vHSD92] P. van Hentenryck, H. Simonis, and M. Dincbas. Constraint satisfaction using constraint logic programming. *Artificial Intelligence*, 58:113–159, 1992.

[Wal94] M. Wallace. Applying constraints for scheduling. In B. Mayoh, E. Tyugu, and J. Penjam, editors, *Constraint Programming: Proceedings 1993 NATO ASI*, NATO Advanced Science Institute Series, pages 161–180, Parnu, Estonia, 1994. Springer.

[Wil76] H. P. Williams. Fourier-Motzkin elimination extension to integer programming. *Journal of Combinatorial Theory*, 21:118–123, 1976.

[Wil90] H. P. Williams. *Model Building in Mathematical Programming.* Wiley–Interscience, third edition, 1990.

[WYO81] T. Watanabe, C. Yasunobu, and M. O'Kuma. Fault diagnosis by mathematical programming. *Transactions of Reliability*, R-30(4):345–352, 1981.

[Zha93] H. Zhang. SATO: a decision procedure for propositional logic. *Association of Automated Reasoning Newsletters*, 22:1–3, March 1993.

Index

W

Z

MIX
Papier aus verantwortungsvollen Quellen
Paper from responsible sources
FSC® C105338

If you have any concerns about our products,
you can contact us on
ProductSafety@springernature.com

In case Publisher is established outside the EU,
the EU authorized representative is:
Springer Nature Customer Service Center GmbH
Europaplatz 3, 69115 Heidelberg, Germany

Printed by Libri Plureos GmbH
in Hamburg, Germany